The Internet Edge

The Internet Edge

Social, Technical, and Legal Challenges for a Networked World

Mark Stefik

The MIT Press
Cambridge, Massachusetts
London, England

First MIT Press paperback edition, 2000

This book was set in Sabon by The MIT Press.

Printed and bound in the United States of America.

Library of Congress Cataloging-in-Publication Data

Stefik, Mark.
The internet edge : social, technical, and legal challenges for a networked world / Mark Stefik.
p. cm.
Includes bibliographical references and index.
ISBN 0-262-19418-X (hardcover : alk. paper), 0-262-69249-X (pb)
1. Internet (Computer network)—Social aspects. 2. Computers and civilization. 3. Information society. I. Title.
HM851.S74 1999
306.4'6—dc21 99-37752
CIP

For Barbara
for the songs she has sung me,
and the dances we have yet to dance

Contents

Acknowledgments

Thank you to my colleagues and friends for suggesting ideas and clarifying my thinking: Bob Anderson, Michelle Q Wang Baldonado, Jean Baronas, Tom Berson, Daniel Bobrow, John Seely Brown, Harvey Brownrout, Stuart Card, Harris Clemes, Lynn Conway, Jeff Crigler, Drew Dean, Rich Domingo, Ken Fishkin, Matt Franklin, Ron Frederick, Rich Gold, Rich Gossweiler, Tom Gruber, Kris Halvorsen, Steve Harrison, Austin Henderson, Tad Hogg, Bernardo Huberman, Peggi Hunt, Kevin Kelly, Don Kimber, John Lamping, Giuliana Lavendel, Joshua Lederberg, Mary Levering, David Levy, Rafik Loutfy, Brian Lyles, Jock Mackinlay, Cathy Marshall, Sanjay Mittal, Geoff Nunberg, Peter Pirolli, Jim Pitkow, Robert Prior, Margaret Jane Radin, Prasad Ram, Ed Richley, Carol Risher, Bill Rosenblatt, Daniel Russell, Bill Schilit, Paul Schneck, Michael Schrage, Jonathan Schull, Hinrich Schütze, Alex Silverman, Bob Shattuck, Steve Smoliar, Bob Spinrad, Peter Sprague, Eric Stefik, Bettie Steiger, Dan Swinehart, Barbara Viglizzo, Dan Wallach, Xin Wang, Roy Want, Mark Weiser, Jonathan Zaremski, and Polle Zellweger.

Special thanks to Carol Risher for making available copies of the AAP newsletter's "Technology Watch" column from 1980 to1985, and to Bettie Steiger, aka TCA002, for providing me with user manuals from the Source from the same period. Thanks to Jeff Crigler, Giuliana Lavendel, Joshua Lederberg, Carol Risher, and Bettie Steiger—all pioneers to the Internet edge—for discussions and stories about the Net, and to Lynn Conway for many interesting conversations about the next edge, although I have not yet done justice to the range of ideas we explored.

Thanks to Harris Clemes for the lessons and practice in psychodrama that inspired the edge theme of the book. Thanks to Matt Buckley of the

Portola Feed Center for his sense of humor and adventure in helping me determine the amount of hay in a haystack. Thank you to the librarians at Xerox PARC's Information Center for chasing down articles, retrieving hard-to-find facts, and suggesting sources. Thanks to Stuart Card, Kris Halvorsen, and John Seely Brown, my managers at Xerox PARC, for acknowledging the value of spending time thinking about the Internet edge and writing this book.

Thanks especially to my colleagues at Xerox PARC for PARC–2000, an intense exercise in reflection, observation, and planning. That exercise, initiated to help PARC plan its research agenda, also provided me with grist for this book as we were engaged in rethinking how our work relates to the emerging future.

Introduction

When I feel it coming on, I unplug the telephone, lock the door, turn off the radio and just give myself to the song until it's finished.
Kim Baker, 1998

Creative times bring many changes. This book is written at such a time. People are preoccupied with the twentieth-first century and talking about how, in "Internet time," what seems like years of change can happen in only a few months. They discuss, enthusiastically or anxiously, innovations such as new kinds of Internet-based businesses for on-line shopping and stock trading, winner-take-all markets, hot new technology companies, failing companies, journalism on the Net, information overload, using the Net for education and making it safe for kids, protecting users from Net hackers, privacy issues, and digital publishing.

This book is about some of the changes taking place in Internet time. It examines what is passing away with the old millennium and what is coming—and may be coming—in the next decade or so. To help us gain perspective, it looks at parallel events from long ago—such as the confusion that occurred when the Erie Canal was built in the early decades of the nineteenth century and the confusion that accompanied development of the phonograph in the early twentieth century. It also looks at the more recent efforts of those I call "early travelers to the Internet edge"—innovators who, in the 1980s, used various new communications technologies to connect us together electronically with e-mail, digital publications, online shopping, and digital news wires.

Chaos in a Microcosm: Digital Music

. . . and a sound arose of endless interchanging melodies woven in harmony that passed beyond hearing into the depths and into the heights.
J. R. R. Tolkien, *The Silmarillion*

Sometimes a technological shift challenges social and legal systems in profound ways. The music industry, for example, is currently facing such a challenge. In the following I consider current events in the music industry as exemplary of the radical changes in our world that are now becoming visible at the Internet edge.

The various cultural regions of the world create wonderfully different kinds of music that travel from place to place, influencing each other and producing new styles in distant lands. This is not a new phenomenon; for many years, new musical styles have evolved, spread around the world, and recombined with other styles to form yet newer styles.

For most of the twentieth century music was distributed mainly through radio broadcasts and recorded media. In the late 1990s, that began to change when the music scene discovered the Internet. A garage band called Severe Tire Damage gave the first live concert over the Net—just beating the Rolling Stones to the punch and, as they put it, annoying people everywhere. Many small, undiscovered bands suddenly realized that they could distribute their music over the Net without the blessings or support of the big labels. These events hinted that something different was about to happen in music distribution.

Over the years, recording companies have maintained their worldwide control of music distribution. Although overseas manufacturers make pirate copies of CDs and there is some domestic copying of cassette tapes, the level of piracy has not threatened the financial foundations of the music publishing business. Since about 1993, however, the price of machines that can reproduce CDs has dropped steadily: from about forty thousand dollars to fifteen thousand, to four thousand; they now cost about two hundred dollars, making it affordable for many people to "write" or copy their own CDs.

In the late 1990s a new digital compression format for music presented an even greater challenge to the music publishing industry. The new format, a form of digital compression for audio files known as MP3, Motion

Pictures Expert Group Audio Layer III, was developed for audio compression. The MP3 format makes it practical to store, play, and transmit musical recordings on computers. Whereas the format used for ordinary CDs requires about ten megabytes per minute of music, MP3 is ten times more efficient. Public interest in MP3 exploded in 1998. It enabled college and high school students with more time than money to distribute their favorite music inexpensively via e-mail and the Net. Today, even a casual search of the Net for MP3 turns up hundreds of sites offering music, players, encoders, and other services.

Portable MP3 and Secure Music Distribution

Ilúvatar said to them: "Behold your Music!" And he showed to them a vision, giving to them sight where before was only hearing, and they saw a new world made visible before them.
J.R.R. Tolkien, *The Silmarillion*

Some people argued that MP3 was no threat to the big labels, because Joe Average does not want to listen to music on his computer anyway. In 1998, however, the MP3 world began buzzing about a new handheld music player by Diamond—The Rio—the MP3 equivalent of a Walkman. The Rio costs around two hundred dollars, can hold hours of music, and downloads music from the Net through a computer. Release of comparable products, such as the Yepp from Samsung and the MPMan from Saehan, were expected in 1999. (See chapter 2, "The Portable Network," for a discussion of the "Cambrian explosion" of innovation in other kinds of portable networked devices.)

When the Rio first hit the market, there was an immediate flurry of legal action. Music publishers and record companies sought an injunction to block sales of the Rio on the grounds that it encouraged unauthorized copying of copyrighted music. A great deal of legal precedent related to copyright protection and the manufacture of copying devices, especially photocopiers, already exists. (The products of Xerox, the company I work for, were central to court decisions on this issue as it related to the Copyright Act of 1976. See chapter 4, "The Bit and the Pendulum," for a broader discussion of the social and legal issues surrounding digital copyright.) Historically, a key issue in deciding the merits of such legal cases has been whether a copying device has other legitimate uses. This criterion was also central to decisions in cases challenging videocassette recorders when they were first introduced. One of the defenses used by makers of the Rio was that it could not itself be used to make copies—a claim that came under scrutiny when two sets of young hackers reportedly succeeded in altering the programming of the device to do exactly that.

Sensing a change in the winds, many technology companies entered the fray. IBM announced its Madison Project infrastructure, which would allow music distributors to sell music over the Net with a lower risk of copyright infringement. IBM possessed many of the components for such an system: high-end servers to store large quantities of digital content, a micropayment technology to allow end users to buy either a song at a time or an entire album, and a digital rights management infrastructure (CryptolopesLive™) to track payments. Nor was IBM alone. AT&T initiated the a2b Music project for delivering songs securely over the Net. InterTrust, a privately held U.S. company, announced a partnership to combine its rights-management and protection technologies with MP3 and advanced audio coding and compression technology from the Fraunhofer Institute for Integrated Circuits. By 1999, Sony Corporation was also ready to enter the market with two separate copy-protection systems—MagicGate and OpenMG—both of which scramble music files before they are sold and unscramble them for listeners. Sony claims its technology allows digital music to be moved, rather than copied, and prevents unauthorized copying, playback, and transmission. (See chapter 3, "Digital Wallets and Copyright Boxes," for a discussion of the real security in such trusted systems.)

By 1999 recording-industry groups were meeting to discuss standards for secure Net delivery of music. At a meeting organized by the Recording Industry Association of America, a hundred companies reportedly agreed to devise a plan to do so. Meanwhile, five major music companies founded the Secure Digital Music Initiative to create a standardized way of distributing music in North America, Japan, and Europe. Leonardo Chiariglione, an Italian researcher who helped create MP3, was named its head.

Digital music distribution is controversial not only because it is currently being used for unauthorized distribution, but also because it has the potential to upset the balance of power in the music industry, enabling recording companies and artists to by-pass record stores and sell directly to consumers. Changes of this magnitude would shake up any industry. (See chapter 7, "The Edge of Chaos," for a discussion of industry dynamics in a time of change.)

Do Audiophiles Want Audio Files?

And in every age there come forth things that are new and have no foretelling.
J. R. R. Tolkien, *The Silmarillion*

What does all this mean for people who are more likely to listen to or buy music than to create or distribute it? Will music distribution really work in an altogether new way? Will we need to log on to our computers in order to buy or listen to music?

Prediction is an uncertain game. Nonetheless, as I look in the following chapters at the grand sweep of changes that are happening now in Internet time, and at shifts that resulted from earlier connection technologies, I sense some contours of the future. One of the enduring properties of successful technologies in the past, and a driving feature of inventions people are now inventing for the Net, is that they provide an increased sense of personal control. (See chapter 10, "Indistinguishable from Magic.") A reliable rule of thumb for judging the potential of a new technology is, therefore, "Convenience rules." If the distribution of digital music follows this rule, we can safely make some predictions about future music distribution.

Given MP3, there are now two main ways to get a personal copy of a musical recording: buy a CD (or cassette tape) or download it from the Net. Although both of these methods will be around for awhile, within ten years I believe they will be secondary to other methods. Consider the following scenarios.

- *"I want that."* You are driving home in your car listening to the radio. You hear some great music and, without taking your hands off the steering wheel, you simply say "I want that." As in some high-end cars equipped with hands-free cellular telephones, your voice is enough to start a transaction. The digitally-enhanced receiver in the car identifies the music and registers your request for it. When you get home, you step up to your home entertainment center, which has by now communicated with the entertainment system in your car. It brings up all the names of music you have asked for recently and lets you listen to samples of them; it also makes available information about the artists and about other recordings of the same pieces. You can choose to buy a copy of the music (for, say, $10); add it to the repertoire of your entertainment system and pay a small fee (say, a quarter) each time you play the music; or purchase the right to unlimited plays for a year (for, say, $5). Given a system of networked appliances and a user design in which convenience rules, you don't need to go to the store or even remember the name of the musician or tune. You simply pay an amount that reflects your degree of interest in the music.

- *Music Club.* In a variant of the first scenario, when you purchase some music you are invited to join an "album of the month" club. Your home entertainment system keeps a profile of your musical tastes; every month you will get an opportunity to listen to and purchase at a discount music from among the hundred or so new albums that fit your profile. Once again, the operating principle is that convenience rules. You don't have to drive to shop, you are kept abreast of new music in your area of interest, and you receive a discount through the club. The publishers win too, because they can rely on fairly level sales from their customer base. In the broader world of publishing, a publisher that develops a business based on subscription sales generally sticks with it because it yields a stable, predictable cash flow.
- *Personalized Radio.* Imagine that you are listening to a radio station in your car or at home. This radio station, however, is specialized to your tastes. It carries the news programs you find most interesting, plays the music you like best, and—most importantly from the perspective of the provider—carries advertisements specially targeted to your taste and buying habits. (See chapter 2, "The Portable Network," for a variant on this idea for reading digital works by means of sponsored browsers.) Technologically, there are many ways to achieve individualized radio. At home, the "radio" channel may be delivered to you via the Net on fiber-optic cable. Your home entertainment system may preload the music onto your car entertainment system. Or it may be programmed to play in between news broadcasts. You do not have to pay for either the music or the news as the advertisers pay royalties in exchange for the time you spend listening to personally targeted advertisements. (See chapter 8, "The Digital Keyhole," for a discussion of related privacy issues on the Net.)

At the time of writing, systems to completely support these scenarios are not available. However, all of the elements for creating them are understood and, given relatively minor extensions of existing technologies, they are quite possible. They represent ways of distributing digital music that are different from the current music business but well within the foreseeable parameters for all types of business at the Internet edge. I believe they will come to dominate the existing modes of music distribution. Convenience rules.

Digital Doom or Information Golden Age?

When hunting, Hawk seeks Mouse . . . and dives directly for it. When hunting, Eagle sees the whole pattern . . . sees movement in the general pattern . . . and dives for the movement, learning only later that it is Mouse. . . . So Hawk—the tendency to look at the Specific—and Eagle—the tendency to look at the Whole—have something to say to one another. . . . each encourages the other toward heightened acuity.
Paula Underwood Spencer, "A Native American Worldview"

At a workshop at the Xerox Palo Alto Research Center (Xerox PARC), where I work, a colleague described how he imagined the Internet would change the world. He believes that we are on the verge of a global transformation in which the individuals in society will become much more interconnected. "Consider your liver," he said, "or the cells in your liver. They are bound together and depend on each other and on other cells in your body for survival. The world is becoming more like that, where our sense of individuality and interdependence is changing."

This statement conveys a sense of wonder of the kind we often attribute to children. Yet the ranks of technology also include many whose optimism has been tempered by intense struggles with technology that does not work very well. What is ahead? Digital doom or an information golden age? As a technologist, much of my creative energy goes into creating new kinds of things. As a person, I see the effects of our human, and often unconscious, actions on each other and on our world.

From the solitary chair inhabited by all writers, I seek perspective. Gaining perspective calls for breathing, taking wing, sharpening perceptions, and reaching for altitude. I have found it crucial to fly with company and have sought out other flyers. One benefit of the rapid pace of change is that there is a wonderful resource—travelers who have thought deeply about the Net and about change. They are not only people who have led the charge, pursuing one technological vision or another, but also those who have found themselves caught in the middle, on a hot seat, in this process.

My fellow edge travelers include heads of high tech companies, librarians at the Library of Congress, copyright attorneys for the music and book industries, technologists in computer companies, and planners in the banking industry.

Many of them have a sense of *déjà vu* about the Internet. They have experienced firsthand the commotion that accompanies trips to the edge and can help us sort through the confusion. They understand that before people use a new technology they need to agree about which version to use. And before they agree and change they will argue. Some arguments carry the day; some don't. Those who have journeyed to the edge have heard some of the arguments before. They bring back the skills of explorers and, sometimes, maps of the territory. Some of these people have offered me their stories and insights, especially those recounted in chapter 7. Yet the journey ahead will surely have more surprises. What came before was just practice.

This book is dedicated to us all—intrepid pioneers, dreamers, couch potatoes, web surfers, computer enthusiasts, hard-boiled engineers, and naysayers—whether dedicated travelers or interested bystanders hijacked on a journey to the Internet edge.

1

The Internet Edge: Change and Connections

An edge . . . marks the limits of who you are and what you imagine yourself capable of. . . . One of the things about an edge is that it represents a really huge identity crisis. On the right side . . . is a new identity. On the left side is an old identity. . . . The edge protects and conserves your old identity.

Arnold Mindell and Amy Mindell, *Riding the Horse Backwards*, 1991

Come to the edge, he said.
They said: We are afraid.
Come to the edge, he said.
They came.
He pushed them . . . and they flew.

Guillaume Apollinaire, *Abre la puerta*

New technologies bring change. When they are fundamental to transportation and communication they affect more than just the mechanics of how we work and live. Because of them, we find ourselves interacting with people from different cultures who hold different values, people who used to seem very far away. We learn to operate in a global space—experiencing cultural clashes and making cultural accommodations. Technological changes lead to cultural changes—especially changes in the social and legal institutions that are now co-evolving with the technology of the Internet. The Net is not, of course, transforming the world all at once, for change is a bumpy process and sometimes involves pushbacks and retreats as well as advances.

Confusion at Edges

One way to understand the changes brought about by the Net is by analogy to individuals negotiating change in their lives. *Process-oriented psychology*—the psychology of change—describes such changes in terms of *edges*. In the process of changing from what we are into what we are becoming we may get stuck at an edge, a point of resistance. Psychologists Arnold and Amy Mindell write about those edges in their books, *Riding the Horse Backwards* (1991) and *Sitting in the Fire* (1995). They characterize a struggle at an edge as an identity crisis.

We can experience edges about trivial matters such as changing a hair style or about important issues like taking on new responsibilities, speaking out, or asking for help. Even though we feel we should speak out or should confront some problem or person, we are afraid that doing so will change our lives. We may recognize that we need to change jobs but hesitate to leave the comfort of the familiar. We know we need help in doing something but are too embarrassed to ask for it. When resistance at an edge asserts itself, we may become temporarily confused and forget what we are trying to do. The motivation for change fades from consciousness, because it is difficult to keep two conflicting points of view in mind at the same time.

Psychologists often observe people approaching an edge and then backing away from it, only to try again later. The urge for the change drives them forward, but resistance to it drives them back. Oscillation—going back and forth—is characteristic behavior for someone working at an edge. It is not always wise to urge other people to cross an edge; they may not yet be ready for the transformation the edge implies. Perhaps they are still practicing the integration steps required for transformation to their new identity.

Collective Edges

Groups can have edges just as individuals have edges. In a group, confusion at an edge usually shows up as lack of agreement about the right course of action. Members may express contradictory and shifting views. Over time, the people in a group may align themselves to make a change, then back off, only to try again later. This phenomenon can be compounded when changes in group membership alter the group dynamics. New members may join as they see themselves as sharing interests and perspectives.

Some edges take a very long time to cross, or may never be crossed at all. For example, Esperanto has achieved worldwide acceptance as a second language, but at a very low level. It trails far behind English and several other languages in the number of speakers. By contrast, the use of nuclear fission-based reactors for generating electricity is an edge we crossed rather quickly, though we may very well cross back again when the costs and risks associated with it have been completely evaluated. For all possible edges, there are forces for the change and forces that push back.

Technology as a Force for Change

I see many of the developments and issues surrounding the Internet as reflections of a collective edge for the "tribes" of our planet. The *Internet edge* is our response to changes occurring in our lives as the world becomes more closely connected through the Net.

The historian Derek de Solla Price (1983) has observed that key changes in technology have often preceded periods of great scientific discovery. He notes, for example, that many of the first elements in the periodic table were identified in the decade after the invention of the storage battery provided a way to isolate elements through electrolysis. In analogous fashion, many advances in astronomy, biology, and medicine were made in the years immediately following the invention of such instruments of observation as the telescope, microscope, and magnetic resonance imager. In the 1980s and 1990s, devices that enhance our ability to read and manipulate nucleotide sequences and genes have led to great advances in the understanding of human biology and disease processes. These understandings have led, in turn, to the development of specific new drugs and treatments.

In the social arena as well, technologies are often credited with causing, or at least enabling and triggering, many dramatic changes. Vivid examples from the history of the past two hundred years include canals, railroads, automobiles and interstate highways, aviation, the telegraph, telephones, radio, television, and communication satellites—all technologies of connection that radically improved means of transportation and communication. These connection technologies shrink distance and bring together people of different communities and cultures. This is why the edge for technologies of connection is often a conflict between global and local values. Such a conflict can evoke resistance, a *pushback*, as people seek stability and attempt to protect the status quo. The pushback thus acts to preserve cultural diversity.

Although a new technology changes things, its effects are not always immediately evident; often its initial form is not at all like the one in which it later appears so influential. At first, too, a new technology is likely to interact with rather than overthrow existing social processes and the installed base of earlier technologies. There can be a long period of co-evolution during which society and technology mutually influence each other.

New technologies inspired by dreams such as the "global village" often start out with only the crudest of implementations. They are promoted by enthusiasts but discounted and criticized by many others. As they are used, they are improved. During an ensuing period of change and confusion—often lasting decades—society and technology co-evolve to create a new order, which is eventually replaced by the next one. The struggle is between the old order and the new, between what society is and what it is becoming.

The Internet is now at this stage of becoming, a period of rapid growth and change. It is still being invented and is characterized by open options, unknown possibilities, confusion, and imperfect technology. Our social structures, cultural assumptions, and legal structures are co-evolving with the Internet.

Recent Journeys to the Internet Edge

From an eagle's eye view, the Internet and World Wide Web are events like many others in a long series of connection technologies. Indeed, in the years just preceding the rapid growth of the Internet, there were many journeys to the edge.

In 1980, Isaac Asimov, the noted science and science fiction writer, spoke to thousands of people at the Plaza Hotel in New York City. He announced that the event they were celebrating marked the dawning of the Information Age; computers, he predicted, would change the world. Electronic mail, digital news services, and on-line reference books would become available to the general public in homes and businesses across America. What was he talking about? It was *not* the Internet as we know it today, and that's actually my point.

The occasion was the launch of a dial-up, on-line computer service called The Source. Red Boucher, then lieutenant governor of Alaska, was among those captivated by the immediate possibilities of connection. He saw in the Source an opportunity for residents of America's largest but most sparsely populated state to become a closely knit electronic community. He wrote a book about it that was electronically distributed on the Source. Later he helped place digital switches and satellite technology in every Alaskan village with more than twenty-five people. Although the Source itself did not spread to homes everywhere as Asimov had predicted, it did later become a part of CompuServe (as recounted in chapter 7).

The Source is just one of many early efforts to create a connection technology like the Internet. People have been working for years to realize a similar underlying dream. Here are two more examples from the 1980s.

In 1986 Brandt Redd was an undergraduate at Brigham Young University when he was recruited by a technologist and dreamer, Curt Allen, to work on a database system for digital publishing. Foreseeing a day when people would get the information they needed quickly through a computer, Redd and Allen founded, with a few others, what later became the Folio Corporation. The firm's articulated vision is "to empower people with information." By the late 1990s Folio was a major technology provider and distributor in electronic reference publishing; it was acquired by Reed Elsevier and then, in 1997, by Open Market.

Between 1981 and 1986 several other companies raced to create a technology to provide informational services and allow customers to transact certain kinds of business from home. The technology was called *videotex*. Communities were wired up with boxes connecting television sets and keyboards in customers' homes to a central computer via telephone. Information such as stock prices and news reports were accessible instantly and electronically. Videotex users could also read and send e-mail, pay bills, and buy airline tickets. About sixty communities in the United States were wired up. Videotex systems delivered some of what they promised and are still used in France as the Minitel system. In the United States, however, they quickly faded. A videotex system called Viewtron built by Knight-Ridder closed in 1986 after losing fifty million dollars.

One way to view the stories about the Source, Folio, videotex, and other early connection technologies is as collective edge work by the communities of the world. The edge we consider in this book is also a leading edge (if not a "bleeding edge") of technology. When the dreamers, inventors, and entrepreneurs among us sense that an invention will enable people to do something new, they organize others, develop the technology, launch a business, and work to realize their dreams. Usually there is a pushback, a signal of resistance: the technology is not ready; society is not ready; the legal system is not ready. Confusion reigns. In effect, the world tries to maintain its balance and diversity and insists on moving at its own pace. It may not be ready for the change. The dreams and experiments are the edge work by means of which the world tries out new things, learns from failures, and backs off to make adjustments and further preparations.

Local Dreaming, Global Effects

The dreams behind the early efforts sketched above are similar to those presently evoked by the Internet. There are many visions of what the Net can become: a universal library where any book is available electronically and instantly to anyone anytime anywhere; an on-line community, where people can stay in touch with friends and neighbors around the world; an electronic democracy, where a vote or a poll on an important issue can be taken immediately; a digital shopping mall, where people can buy unusual goods at great prices in specialized shops all over the world.

Those dreams give us glimpses into possible new orders: worlds of greater connectivity, knowledge, and experience. But life is complicated; changes in technology lead to changes in society. Change is difficult, and rapid change is even more difficult. When something is new, we often do not understand the problems it presents until we try it, experience the difficulties, and take time to reflect on them.

As awareness of the Internet increases and we approach the new millennium, dreams of world connectivity are entering the public consciousness. We can see that some of these dreams make good sense. Other, still unrealized dreams may ultimately make perfect sense, but something is not yet ready for them.

Gore's Highway and Clinton's Ditch

The probe-and-pushback pattern of collective edge behaviors that surround the introduction of a new technology is clearly reflected in the historical record. In hope and confusion, each new technology appears before it has been perfected. Advocates arise, citing economic and social benefits and urging action. Opponents arise, citing not only technological barriers but also legal and social issues that must be addressed before the technology can be deployed. Change is bumpy, coming in breakthroughs and setbacks. The outcome hangs in the balance while forces for change wrestle with forces for the status quo. This chaotic initial pattern has been particularly evident in technologies of connection.

We can perhaps understand the uneven pattern of early Internet development by looking at similar events in the past. Let us, for example, compare the public speeches given about two proposed new technologies and consider how the arguments used reflect beliefs about anticipated changes. The examples are drawn from public speeches about the Internet in the early 1990s and about the proposed Erie Canal in the early 1800s.

In his book *Erie Water West* (1990), Ronald E. Shaw characterizes the Erie Canal as the work of a remarkable generation of Americans who made the period between 1815 and 1860 an age of great national expansion. In New York this generation proposed to build a waterway that would expand the economy of the state beyond the eastern areas dominated by the old Knickerbocker society; these dreamers sought to penetrate the frontier and to create a new mold for the future of the Old Northwest Territory. Futurists such as Alvin Toffler see the Internet as a corresponding symbol of another generation, one ushering in a new age variously called the information age, the communication age, or the knowledge economy.

In 1993, Vice President Al Gore emerged as a prominent advocate of the Internet. At the time the Internet was still mysterious to most people. The Web had not yet spread out over the world; there were no web browsers, and the wars between the Netscape Navigator, Microsoft's Internet Explorer, and others had not begun. It was several years before network addresses became trendy and useful additions to advertisements and business cards; before "Net Days" were held to connect schools to the Web; before the mixture of push and pull approaches to sending information began to define and redefine the uses of the Net. Gore's speech to the National Press Club that year was widely quoted and popularized the phrase "information superhighway" as a metaphor for the Net.

> It used to be that nations were more or less successful in their competition with other nations depending upon the kind of transportation infrastructure they had.

Nations with deep water ports did better than nations unable to exploit the technology of ocean transportation. After World War II, when tens of millions of American families bought automobiles, we found our network of two-lane highways completely inadequate. We built a network of interstate highways. . . .

Today, commerce rolls not just on asphalt highways but along information highways. And tens of millions of American families and businesses now use computers and find that the two-lane information pathways built for telephone service are no longer adequate. . . . One helpful way is to think of the National Information Infrastructure as a network of highways—much like the Interstates begun in the fifties. (Gore 1993)

Gore went on to predict that the Net would bring prosperity, jobs, new products, and a surge in American exports. But he also foresaw foreign competition and the need for timely investment.

This kind of growth will create thousands of jobs in the communications industry. But the biggest impact may be in other industrial sectors where those technologies will help American companies compete better and smarter in the global economy. Today, more than ever, businesses run on information. A fast, flexible information network is as essential to manufacturing as steel and plastic. . . . Virtually every business and consumer in America will benefit dramatically from the telecommunications revolution. . . . If we do not move decisively to ensure that America has the information infrastructure we need every business and consumer in America will suffer.

Gore's speech follows a familiar pattern of exhortation that can be heard in political speeches advocating earlier connection technologies. In the early 1800s DeWitt Clinton—nephew of New York Governor George Clinton, state and federal legislator, mayor of New York, and himself a future governor—was a land speculator, naturalist, and educator. He was also an ascendant leader of the Republicans in the New York State Senate and a vocal proponent of the Erie Canal. Among his many speeches and "memorials" advocating construction of the canal was a letter to the New York Senate; it was published in 1821 as *The Canal Policy of the State of New-York: Delineated in a Letter to Robert Troup, Esquire.*

The importance of the Hudson river to the old settled parts of the state may be observed in the immense wealth which is daily borne on its waters, in the flourishing villages and cities on its banks . . . if . . . this great river was exhausted of its waters, where then would be the abundance of our markets, the prosperity of our farmers, the wealth of our merchants? . . . what blessing might not be expected if it were extended 300 miles through the most fertile country in the universe, and united with the great seas of the west! (Quoted in Walker and Walker 1963, page 3)

Clinton urged the legislature to appropriate funds for building the canal to Lake Erie to thwart British plans to construct a Canadian canal linking the St. Lawrence River to the western Great Lakes by way of Lake Ontario.

> If a canal is cut around the falls of Niagara, and no countervailing nor counteracting system is adopted in relation to Lake Erie, the commerce of the west is lost to us forever. . . . The British government are fully aware of this, and are now taking the most active measures to facilitate the passage down the St. Lawrence. (ibid.)

The Erie Canal was built between 1817 and 1825. Through a series of locks, it connected the port of New York City via the Hudson River with Lake Erie at Buffalo, so providing a navigational link to the other Great Lakes. The canal contributed to the rise of New York as a major port and facilitated the settlement of Minnesota, Illinois, Wisconsin, and Michigan by immigrants. The people of the Great Lakes states shipped grain east and received manufactured goods in exchange. DeWitt Clinton was elected governor of New York during the building of the canal, which was often called "Governor Clinton's ditch."

A key argument raised for any connection technology is its positive effect on business. Throughout his Press Club address, Gore described the free-enterprise aspects of the Clinton administration's Internet vision in terms of universal accessibility to information.

> This Administration intends to create an environment that stimulates a private system of free-flowing information conduits. . . .It will involve a variety of affordable and innovative appliances and products giving individuals and public institutions the best possible opportunity to be both information customers and providers. . . . Anyone who wants to form a business to deliver information will have the means of reaching customers. And any person who wants information will be able to choose among competing information providers, at reasonable prices. . . . That's what the future will look like—say, in ten or fifteen years. (Gore 1993)

In promoting the Erie Canal, DeWitt Clinton also drew extensively on economic arguments, often from the essays of Jesse Hawley, which were published in the Genesee (N.Y.) *Messenger* in 1807 and 1808. Hawley had predicted the extensive development of the western part of New York State once the canal became a reality. Buffalo, Albany, and all the cities along the canal would prosper. Lock tenders would be paid for by taxes on water-powered mills built along the canal. To handle the increased flow of trade with the western states, New York City would expand to cover all of Manhattan.

The high-flown rhetoric of the time tied the identity and future of New York firmly to canal building. For example, Charles Haines of the New York Corresponding Association for the Promotion of Internal Improvements wrote in 1818 that "the state of New-York will never rest, till she sees the waters of the Lakes mingle with ocean that washes her coasts. Her interest, her pride, her glory demand it. Her faith is pledged, her will is spoken, her arm is put forth, and who shall arrest it!" (quoted in Walker and Walker 1963, page 19).

Many proponents emphasized that the ultimate success of the settlements in the upper Midwest was irrevocably linked to the canal. Taking a similar tack almost two centuries later, Gore argued that the benefits of the Internet would spread far beyond business: "But, even more, as I said at the outset, these methods of communication allow us to build a society that is healthier, more prosperous, and better educated. They will allow us to strengthen the bonds of community and to build new 'information communities'" (Gore 1993).

Of course, when Gore and others spoke about the Net, their proposals, like earlier plans for building the Erie Canal, were not universally accepted. When representatives from New York traveled to Washington seeking federal funds to build the canal, they were rebuffed by no less a man than Thomas Jefferson. Hosack's *Memoir of DeWitt Clinton* quotes President Jefferson's response to their request.

> Why sir, here is a canal for a few miles, projected by George Washington, which if completed, would render this fine commercial city, which has languished for many years because the small sum of two hundred thousand dollars necessary to complete it, cannot be obtained from the general government, the state government, or from individuals—and you talk of making a canal 350 miles through the wilderness—it is little short of madness to think of it at this day. (Quoted in Shaw 1990, page 36)

In general, arguments against the Erie Canal centered around expense and distance. Opponents said that the distance and the costs were too great and that the taxes levied to build the canal would be unfairly borne by citizens least likely to benefit from it. Many of the predictions made by people on both sides of the issue turned out to be correct. There was as much confusion about the future then as there is now.

The Change Amplifier

Nothing changes alone. When something changes, other things are affected. When a company opens in a town, new businesses and new jobs are created. Houses get built. Stores are expanded. New people meet each other. In ways that no one can predict, kids become friends with kids from other places and lives move in new directions. These changes are part of the way a town evolves and adjusts to new situations. If a company closes down, the opposite effects may occur. Jobs are lost. People move away. The chain of causes and effects may lead back to company leaders at a corporate office far away or to a slight change in the price of a material in a distant part of the world.

The effects are not always predictable. Change brings chaos. A small change can trigger unexpectedly large changes in lives far removed from the point of decision. From a philosophical point of view, everything is impermanent and change never stops. Things are always changing in a cycle that breeds further changes. There is always chaos somewhere. Nonetheless, some things endure through periods of change long enough to let us become accustomed to stable conditions. We notice when there are big changes.

The Internet amplifies change. Like all earlier connection technologies, it does so by reducing the power of distance. People say that the world is shrinking. Of course, the planet hasn't changed in size, and great distances can still limit the pace of change. For the most part, it is still true that the farther apart things are the less they interact. In basic physics the law of gravity says that the gravitational attraction between two objects decreases with the square of their distance from each other. Something analogous can be said of social processes. In an office building, one of the major factors governing how often people talk to each other is how far apart their offices are. The perceived "costs" of walking to someone else's office goes up with its distance from our own. We recognize intuitively that a work group is more effective when its members' offices are close together.

Distance in these examples is not simply straight-line or Euclidean distance. When people have to go up or down stairs—working against gravity—to talk to each other, the actual reduction in interactions is even greater than when their offices are at more distant points on the same floor. When there are walls, locked doors, or other barriers in the way, the effort required to get together becomes part of the perceived "cost" of the interaction.

The Internet lowers such costs; it doesn't eliminate them, but it does foster more action at a distance, so that something happening over here can have an effect over there. The fan-out effect of the Net can cause multiple changes at many distant locations. And, because each change triggers further changes, the pace of change accelerates and the potential for chaos increases.

The ability of the Internet to cause action at a distance is akin to that of other communication technologies—including smoke signals, drum talk, mail, newspapers, radio, television, and the telephone. It enormously magnifies the most powerful feature of technologies like radio and television: wide dissemination of information. Its enormous fan-out allows an idea or an action originated at one location to be sent instantaneously to thousands or millions of other sites. In addition, like the telephone, the Internet is interactive. Some have described Internet technology in terms of "push and pull." Imagine a playground full of kids playing tug-of-war with long ropes, pushing and pulling, yanking and shoving, creating interaction at a distance.

Barriers and Change

Barriers protect us from too much outside influence. Fences are barriers. Locked doors are barriers. Walls are barriers. "Good fences make good neighbors," the comment attributed to his neighbor in Robert Frost's poem *Mending Wall,* reflects our perception that barriers contain or reduce interactions and guard against possible friction.

In nature, rivers and mountains can act as barriers to the migration of animal and plant species. Great distances, such as the distance across the great Australian desert, act as similar barriers and play an important part in creating and preserving biological diversity. Separated by a barrier, populations of life forms stop interacting. The populations on each side become subject to different evolutionary pressures. Chance mutations, which in the right conditions may become established in the gene pool on one side of the barrier, may not occur on the other side. Over time, populations of separated life forms that were initially a single population develop distinctly different gene pools. When they become so different that they can no longer interbreed, they are classified as different species.

An analogous mechanism involving the separation of populations and independent evolution can help explain the diversity of human cultures and languages. Groups of people move away from each other, and over time their language patterns diverge. After a little drift, they speak different dialects. The French spoken in Montreal is not the French spoken in Paris; the English spoken in Australia is not the English spoken in England or the United States. When the separation lasts long enough and interactions between the populations are few, distinctly different languages evolve. On the other hand, a contrasting effect can be heard in the French spoken in Toronto, which is said to be essentially "Parisian"—because it is being constantly refreshed by education and new immigration from France.

For many countries, national borders still constitute cultural and linguistic barriers, reducing outside influences and inhibiting interactions. Similarly, when countries raise or lower tariffs on trade, they are trying to control the rate of economic influence from abroad. High tariffs, referred to as "trade barriers," shelter local interests from outside forces that could bring about change.

It is said that national borders are "just speed bumps on the information superhighway." The comment reflects our realization that in the 1990s and beyond the Internet connects computers on either side of a national border with the same ease as it connects computers in the same city. Information packets, e-mail, and digital money zip across borders without going through customs, export control offices, or any entity that reflects national economic interests. As the effects of borders are reduced, a country's cultural life is equally subject to influence by actions taken elsewhere in the world.

Barriers in Cyberspace

From the foregoing discussion, we might suppose that because the Internet easily crosses physical borders, there are no barriers within cyberspace. That is not the case at all.

Consider an on-line chat room. People in different parts of the country or world can log on to an on-line service provider and search through the topics or discussion groups active at a given time. They select a group, and their computer connects them to the ongoing discussion. The chat room creates a sense of place because all the people "in" the chat room can communicate with each other in real time, no matter where they happen to be physically. A conversation in a chat room is akin to a meeting in a real conference room that is separate from other rooms in an office building.

Physical distance is irrelevant to chat rooms. People thousands of miles apart can interact in a chat room as a closed group, while two people seated in the same physical room may be participating in two completely different chat rooms. Yet, if we consider the memory cells of the server's computer as the two people send and receive messages in their respective chat rooms, one room is probably less than a centimeter away from the other, perhaps on a different portion of the same chip. The barrier that keeps the conversations in different chat rooms separate is determined by programming rather than by physical distance.

Chat rooms are not the only Internet phenomenon in which barriers are deliberately created. In an effort to keep their private networks (*intranets*) safe from intruders, many organizations mount so-called firewalls—programmed gateways that monitor and control traffic between two networks. Depending on their programming, the gateways allow some kinds of packets through and block others.

There can even be barriers to objects that move freely around on the Net. For example, one approach to protecting digital documents is to put them in encrypted "digital envelopes." Such envelopes are designed to allow access only to parties or systems that possess appropriate digital authorizations and decryption keys. This is analogous to passing things around in the real world in locked boxes. The boxes are portable, and only people with keys can look inside.

Thus barriers within cyberspace—separate chat rooms, intranet gateways, digital envelopes, and other systems to limit access—resemble the effects of national borders, physical boundaries, and distance. Programming determines which people can access which digital objects and which digital objects can interact with other digital objects. How such programming regulates human interactions—and thus modulates change—depends on the choices made. In the late 1990s, these choices are mostly in the hands of such relatively ad hoc technical bodies as the IETF (Internet Engineering Task Force). I am not suggesting that such groups are exercising that power improperly but, rather, that as the Net increases in its ability to affect action at a distance—especially economic action—political interests are likely to seek influence over those choices.

Reflections

Working between the interplay of change and pushback at the Internet edge are the people and organizations directly concerned with the development and deployment of technology. Engagement at this hot point of change tends to foster a perspective shaped mainly by the need to solve economic and technical problems. At the Bandwidth Forum sponsored by the Federal Communications Commission (FCC) in January 1997, Stagg Newman of Bellcore addressed the question of the design of the National Information Infrastructure (NIII) in those terms: "In the early 1990s there was lots of debate over what the NII should be. . . I think the Internet has answered that for us. NII will be IP (Internet Protocol). . . . And the key to really unleash the potential is having a broad band" (Newman 1997).

Newman's comment reflects his own reputation as a talented technologist at a major communications company. His reference to "what the NII should be" is intriguing, in part because it is a question that could be answered at several different levels. Newman, however, does not risk becoming lost in metaphor; he focuses on technology. For him, the NII is defined by the IP layer—that is, the protocol originally developed by Robert Kahn and Vinton Cerf for the packet-switched ARPANet and now used throughout the Internet. It is the key to creating a whole gamut of digital services; everything important from a technological perspective can be built on top of that protocol. What matters the most in realizing the Net's potential, Newman argues, is providing adequate bandwidth. He then goes on to consider various options for achieving efficient digital communications.

In several technologically advanced nations—including the United States, Japan, and a few others—engineers are working to lower the costs of digital networking, systems, and services. Companies are scrambling for position as specialists in different areas set about solving a cluster of technical problems involving bandwidth, wiring, encryption, displays, electronic commerce, batteries, and other system components. As costs drop and bugs are worked out, the technology that started in a few technologically advanced countries is spreading over the world. Simply put, technology drives change. The world is going digital. It is at the tip of a wedge of great change.

But there is another way of looking at "what the NII should be." Culture and dreams shape the conscious and unconscious view of economics, and this drives the development of technology. Simply put, culture and dreams drive technology. They also shape the pushback from the Internet edge.

Although the Net has already spread to many parts of the world, it is evolving most rapidly in the United States, where almost all sectors of society are responding to its influence. That response makes the U.S. a good place to sample the global Net of the future and to consider issues that go beyond economics and technology. This is not to say that what works in the United States will work everywhere in the world; nor that systems and technologies developed elsewhere will not affect U.S. users. The point is that the levels of development and evolution now seem highest in the United States.

The journey we are all on involves uncertainty, prediction, and choice. While the Internet is being created we live in the chaos at the Internet edge, where the technological and economics perspectives are not the only ones that matter. We experience the tension between local and global values as the world becomes more connected. We dream; we design; we build or participate; we see effects; we reflect; we make choices; we repeat. We are living the dream at the edge of transformation.

2

The Portable Network: Away from the Desktop and into the World

No matter how much PC vendors want to convince us that sitting in front of a computer monitor twenty-four hours a day is the ultimate fulfillment, it simply will not happen. [. . . there are far too many places that digital media can be put to good use for it to remain tethered to the desktop.]

Denise Caruso, "Digital Commerce"

The early history of the phonograph is a story of competing inventors, competing advances in technology, and competing visions of the nascent market for sound recording. In 1877, Charles Cros in France and Thomas Edison in the United States both described practical methods of recording sound waves on a disc or cylinder and playing them back. Cros developed no working model, but Edison had his machinist, John Kruesi, build a device in November of the same year. In 1878 Edison received a patent for both disc and cylinder formats. His first models used a sheet of tinfoil wrapped around a grooved metal cylinder. As he spoke into a horn, the sound waves made a diaphragm vibrate, causing a needle to record the vibrations as indentations on the foil as the cylinder turned.

In 1881, Charles Sumner Tainter and Chichester Bell introduced the wax cylinder, which had superior recording capabilities and could be reused by simply shaving off the layer of wax holding the earlier recording. Their work at the Volta Laboratory was sponsored by Alexander Graham Bell, who was interested in adapting his telephone invention for remote dictation. They called the device the graphophone, received a patent on it in 1886, and demonstrated it at a major exhibition in Washington, D.C., in 1889. They subsequently sold the patent to the American Graphophone Co.

In 1887, Emile Berliner patented what he called a *gramophone,* which engraved sound waves in a spiral groove on the surface of a flat disc. He demonstrated it the following year. In 1897 he founded the Berliner Gram-o-phone Company in Montreal and, with his brother Joseph, the German company, Deutsche Grammophon.

Where an Invention Ends Up Is Not Always Where It Starts Out

What did these inventors think sound recording would be used for? Even though Edison received an enthusiastic reception when he played recordings of music and comedy in a public hall, he believed that the major use of his invention would be recording dictation in business offices. The sound quality of his early apparatus, however, was barely good enough to reproduce intelligible speech. Bell and Tainter also expected dictation to be the primary application for sound recording.

Emile Berliner, however, saw the future more clearly, with the entertainment business as the major market for sound recording. Even though the sound quality of Berliner's discs was initially poorer than that of Edison's cylinders, his invention of a method to mass-produce recorded discs at low cost made the subsequent development of the recording industry possible. Moreover, Berliner's motor supplier, Eldridge Johnson, who took over the patent when Berliner was forced out of business, greatly improved the sound quality of the flat disc. In 1901 Johnson used the gramophone patent to form the Victor Talking Machine Company. Thereafter the superior sound quality and durability of the flat disc steadily attracted performers and the public. By 1910 discs recorded at seventy-eight revolutions per minute (78 rpm) had become the dominant format.

In the late 1890s, however, technical shortcomings and bitter patent battles had impeded the growth of the phonograph business. In the years between 1896 and 1901 the competing patentholders consolidated and regionalized the industry. The North American Phonograph Company purchased controlling interests in both Edison's and Tainter and Bell's patents. Its primary purpose, at first, was to manufacture office dictating machines.

At the turn of the century, the main companies in the market for sound recordings included Edison's National Phonograph Company (successor to North American Phonograph) and the Victor Talking Machine Company. In the 1890s North American Phonograph had begun to lease regional rights to subsidiaries doing business under different names. Several of the subsidiaries began expanding into the entertainment business. One of these lessees, the Columbia Phonograph Company, operated in the Baltimore and Washington, D.C., market. By the end of the century Columbia had a catalog of over five thousand recordings, including popular songs, military marches by John Philip Sousa, political speeches, and various novelties. Indeed, a primary way of distributing political information in this period was to play recorded speeches from the stages of public auditoriums. For the first time, large numbers of citizens became acquainted with the distinctive voices of politicians and orators such as Theodore Roosevelt and William Jennings Bryan.

Columbia later became Columbia Graphophone, which, in 1907, split into a business segment (which later became the Dictaphone Corporation) and an entertainment company (which later became Columbia Records). Columbia had been offering flat discs since 1901 and in 1912 ceased manufacturing cylinders altogether.

Looking over this brief early history of the phonograph, we are struck by how the initial visions of the inventors and their companies shaped and limited the opportunities they pursued. Although sound recording is useful for office work and business dictation, its value does not lie primarily in that application. Berliner's goal was to make recordings for mass entertainment, and his and Johnson's work on sound quality and mass production followed from this goal. Although the rest of the industry eventually decided to move in the same direction, Berliner's sustained focus enabled him and his successors and collaborators to catch up from behind, set the standards, and dominate the market. With their attention focused on developing dictation machines and their energies sapped by patent disputes, the companies manufacturing machines invented by Edison, Tainter, and Bell squandered their early lead.

What Is the Net Good For?

Today the Internet is in an analogous period of new invention, competing visions, advancing technology, and companies scrambling for advantage. The Net has the potential to become many things, and people have different views of what it is good for. Like the early phonograph companies, today's Internet companies are likely to find that their goals and visions will limit and shape their opportunities.

What comes to mind when you think of someone using the Net? Many people first visualize a "net worker" as someone sitting in an office in front of a display terminal and typing on a keyboard. Such thinking about the Net is mainstream, but it is too narrow. The Net has a larger place in the world than the desktop.

The real power of the phonograph could not be realized until people got beyond the idea of using it only for office dictation. In the same way, the real power of the Net will not be realized until we think beyond office work. To understand what the Net can be we need to both expand and contract our range of vision. We need to think big, think small, think microscopic. Think built-in, think portable, think wearable. Think inside, think outside, think ubiquitous. Think mornings, think evenings, think continuous. Think getting information from the Net, think sending information to the Net. Think specialized, think universal.

So what could the Net become, besides an adjunct to office work?

• *The Big Screen:* A big screen for digital news and entertainment. Lounging on a comfortable sofa, you could watch and interact with the presentation from across the room.

• *The Personal Document Reader:* An interactive, lightweight, and portable digital display with two-way wireless communication. From an easy chair at home or anywhere else, you could watch the digital news and read digital articles while holding the PDR in your lap.

• *The Web Watch:* A screen on your wrist, like the videophone in old Dick Tracy comics. Only it's not just a videophone. You could read and send e-mail, check your calendar, and download a client's picture on the small screen.

• *The Web Tricorder:* A palm-sized computer with a digital eye or a digital camcorder with a net connection. It could record digital pictures or videos and post them on the Net in real time via a wireless connection. It could also retrieve and display information and video from the Net.

• *The Insighter:* A discreet handheld device that could scan a business card, search an information service on the Net, and summarize crucial information about a business contact on its small screen.

• *The Hiker's Companion:* A pocket-sized information device combining global positioning with on-line information service. The wanderer in mountain or forest could keep informed about trails and events back at camp and summon help in trailside emergencies.

• *The Net Phone:* A compact cell phone handset that could also serve up the interactive web on a miniscreen.

• *The NetMan:* A sporty yellow headset with earphones and a stem microphone. You could use it to listen to music on an early morning run or interact by voice command with e-mail, your calendar, or Net information sources.
• *The StreetSmart:* A combination Net-based information kiosk and videophone built into lamp posts downtown. You could use it to get directions at a street corner or to find out about nearby stores and restaurants.
• *The Web Suit:* Radio-linked wearable appliances that sense the environment, provide information through glasses and earphones, and interact with the user through forearm keyboards or voice commands. The suit would keep the wearer informed of events in his or her environment or on the Net and provide digital controls for browsing or controlling other networked devices. (See chapter 10.)

By the late 1990s, all the functions in these examples were available in some form on the Internet. From our desktop we can see and record video, read news, check e-mail, consult calendars, get directions, telephone friends, and trade stocks. Each of these capabilities changes our view of the Net in a different way. For example, they alter our perceptions of the most advantageous size and shape for the computer. Even though we can watch a movie at our desktop, we'd far rather lounge in a comfortable chair. Even if we can read or watch the news at our desktop, we generally use our desks for different purposes. We do not carry a big computer on the road. Nor, if we are walking, do we want to type. For mobility we need something about the size of a telephone that we can interact with vocally while keeping our hands free.

The Digital Cambrian Period

Bill Buxton (1997), an expert on user interfaces and sometime musician, argues that general purpose computer systems and interfaces cannot deliver their intended functionality. When too many kinds of programs are available on an interface it becomes too complex to use. In his view, the device itself should become the interface. Consider preparing to take a photograph with a camera that has a through-the-lens viewer. You turn the camera toward a scene and the scene comes into view in the viewfinder. You turn to the right, and the camera shows the view on the right. You pull the lens in and out (or turn it), and the scene comes closer or backs away. These motions are the *affordances* of the camera; they support the choices that a photographer makes in setting up a picture. Other elements, like adjustments to lighting and film speed, have more obscure controls. In a well-designed device like a camera the affordances reflect the requirements of the task. For Buxton, the enemy in interface and system design is not generality but, rather, generality at the expense of complexity.

To restate the argument: We do many different things, such as taking pictures, reading mail, talking to others, and looking for information. Many of these things could be wonderfully augmented or amplified by a networked device. But each task has different requirements. Trying to put all of the functions into one multipurpose device makes using it too complicated. What we need is a family of specialized, networked devices that can work together, each device designed to perform one task effectively and easily. As the Net becomes ubiquitous, therefore, we should stop looking for the perfect Internet device. The Net can become part of the context for all devices. It is the infrastructure for communication, information, control, and, especially, connection. As the slogan for the 1997 World Wide Web Conference says, "Everyone • everything • connected."

Early versions of many of the Net devices mentioned above are already being developed. Web-TV, laptop and pen-based computers, digital cameras, GSM telephones (global system for mobile communications), videotex information booths, and handheld communicating devices are all available. The current period of rapid invention and change is almost like the Cambrian Era, a period of expansive genetic diversity about 544 million years ago, when almost all the major phyla of animals first appeared. Animals burst forth in a rush of evolutionary activity never since equaled. Ocean creatures acquired the ability to grow hard shells and a broad range of new body plans emerged.

Similarly, in the late 1990s, new digital and networked devices appeared every week, challenging our understanding of both the possibilities and the categories. There are desk and cellular telephones with built-in computers that can dial-up your e-mail account, and there are computers with built-in telephones that can support digitized talk over the Net. There are digital cameras for desktop computers, palmtop computers with built-in cameras, and cameras with built-in computers that become shareable video servers when plugged into the Net. You can make a video "telephone call" on a personal computer without using a telephone or make a call by way of television without using a personal computer. Communicating flat-panel computers come in sizes corresponding to the business card, the pocket calculator, the clipboard, and the blackboard, and manufacturers are experimenting with sizes in between. In the run-up to every holiday season and technology show, creative people around the Net search for new products they hope will be popular. By the next show many of last year's offerings have been tossed aside and forgotten. There have been pushbacks from the Internet edge, usually because some aspect of the new creations wasn't quite right.

Thus, the Internet isn't turning out to be just something that connects to the computers in our offices. It is already developing a myriad of simultaneous, interconnected forms.

The Wonderfulness of Paper

Paper . . . provides the highest resolution and truest color images on the cheapest, lowest weight display. It allows for annotation, easy disposal and recycling, and for dynamic modification of its form factor—that is, it can be folded and put in your shirt pocket at a moment's notice.

Stu Card, 1997

In the digital age it may seem quite retrograde to praise the virtues of paper. Although papyrus came into use earlier, in around 3500 B.C., the invention of paper is generally dated at about 105 A.D. and attributed to Tsai Lun, a Chinese court official. Paper and papyrus were the medium of choice for creating a permanent record and for preserving thoughts and ideas as well as records of agreements, commercial transactions, taxes, or family lineages.

Paper is wonderful. It has high contrast. It can be written and drawn and printed on with colorful inks and used to reproduce photographs. Paper makes written ideas both durable and portable. The wonderfulness of paper is a big part of the success story of Xerox, where I work. When photocopiers were first invented, nobody was sure exactly what they would be used for. According to Xerox Corporation lore, Chester Carlton, the inventor of Xerography, was unable to interest big companies in selling photocopiers, because no one believed there was a market for them. Prior to the photocopier, the most common means of copying a document were carbon paper and mimeograph machines. By now those older technologies have been so thoroughly displaced that many children born in the United States since 1980 do not know what they are.

When photocopying became routine, people could show up at a meeting with notes or an agenda and copies for everyone. Attendees could take these copies away with them and make more copies for others—spreading ideas more widely and rapidly within and without organizations. Scholars at libraries could easily copy sections of books and other records, making scholarship more portable. Although such uses of paper and photocopying are now easy and ubiquitous, in historical terms they are a very recent development.

When multimedia systems appeared on personal computers in the early 1990s, some people said that paper was outmoded, because it is a passive medium. The use of paper for documents was expected to fade quickly. Multimedia is flashy. Like television, it catches our attention with action, interactivity, and sound. Kids raised on television are immediately drawn to it. Most children, argued the prophets, don't read; they watch and interact. Paper documents were doomed and would be superseded by multimedia documents.

Yet paper is still used, because there was a pushback from the edge. The resistance was felt as skepticism by the not-yet-ready and as frustration by the multimedia enthusiasts. The debate about multimedia versus paper—reminiscent of the discussions of prospects for a "paperless office" in the 1970s and 1980s or about the "future of the book" in the 1990s—are a prelude to understanding what is ahead. Although new forms of documents and communications have advantages, it is clear that paper and passive documents have staying power. In 1997, well over a trillion pages were generated by printers and copiers in the United States. Given the fact that computer screens have not replaced paper for reading, what is the nature of the struggle at this edge?

We can start by asking, "What do computers (or the new documents they generate) have that paper lacks?" To begin with, multimedia documents are lively; and, because they are interactive, they can deliver different experiences and information to different people. Digital documents in general can be delivered quickly and electronically. A digital stock market report can be updated continuously, or at frequent intervals. With their animations, interactivity, and rapid delivery, multimedia documents capture the pulse of a society on the move.

Why then does paper survive? What do paper documents have that digital documents do not? To publishers of books, newspapers, and magazines, their most important property is an established economy for buying and selling them. (I discuss this issue in relation to digital publishing in chapter 3.) Other properties of paper documents are so obvious that we may underestimate their importance: they are portable, cheap to make, usually don't break in ways that affect their readability. Their visual presentation has good color and contrast. They require no electronic devices for reading them. Consequently, human readers need not contend with low batteries, out-of-date formats, or failures of the operating system. Paper can also be wonderfully personalized. You can write it, fold it, attach things to it, color it, copy it, bind it together in a package. It has no computer program to say "you can't do that."

An Internet Dream: Information on the Go

What would it take to bring the portability and other properties of paper documents to devices for the Net? In the sample list of untethered visions for the Net, the one that most directly addresses this question is the portable document reader (PDR). In time (perhaps not very much), PDRs may become the new digital paper, just as useful for reading and marking things up. At their best they could combine the interactivity of computers, the connectivity of telephones, the bandwidth (the rate at which data can be transmitted or received) of television, and the portability of paper. The search for the perfect PDR has been a particularly enduring and compelling dream. Looking at the advantages of PDRs and the obstacles to their development will help us understand how technology and the Net are co-evolving.

The Technology of Portable Document Readers

Many technologists, librarians, students, and futurists I met during the 1990s liked to describe their ideal version of a PDR. Usually they modeled it after a tablet or pen-based computer. It would have a bright, high-contrast display area about as large as a letter-size sheet (8½ × 11 inches)—or, for Europeans, an A4 sheet (about 8¼ × 11¾). It would weigh less than a pound and have enough storage capacity to contain several complete works, if not a small library. When used as a portable, the batteries would last for many hours, if not days or weeks.

For many, an important consideration is ease of reading. Because people often imagine using the device while seated in an easy chair, they usually think it shouldn't have a keyboard. They have various ideas about how to interact with it. Some suggested communication by hand gestures or hardware buttons or touch-sensitive screen buttons for operations such as flipping pages, skipping to a section, highlighting passages, or leaving a bookmark. Over time their proposals for downloading documents to the device have also differed. In the early 1990s, the idea was to load books on floppies; later people suggested that on-line books would be loaded by using a modem. Still later, the idea was to use a built-in cell phone or other wireless technology for cordless and ubiquitous downloading.

Not Quite PDRs

Although laptop computers were the most prominent portable computer products available in the 1990s, they are decidedly less "bookish" than PDRs, because keyboards figure so prominently as the user interface. Nonetheless, the names of these devices allude to their paperlike and booklike properties. In the 1970s, Alan Kay's vision of a laptop computer—complete with fold-down screen, keyboard, audio, and a video camera—was called the Dynabook. Apple, Hewlett-Packard, and Sun Microsystems later named their laptop computer products Powerbook™, Omnibook™, and SPARCbook™, respectively. Handheld devices by Franklin Books are loaded with dictionaries, bibles, and reference books. Although the displays on these devices were much smaller than typical book pages, they introduced many people to the idea of reading on portable electronic devices. In the early 1990s, Voyager began publishing several book titles on floppies with interfaces for reading them on laptops.

In the late 1980s, Wang Laboratories demonstrated pen-based interfaces for their Freestyle™ system, which ran on their proprietary line of desktop personal computers. The interfaces interacted with computers and enabled users to mark up "electronic paper" documents. Implemented on the available CRT-based displays with tethered touch-sensitive sheets and pens, the systems lacked the pen-and-paper feel that later flat-panel systems sought to create and were not portable.

During this same period, other dreamers of the PDR dream tried to build devices that were not only pen-based but also as flat and as portable as paper. Tony Fidler of Knight-Ridder traveled around in 1993 with a mock-up of such a device and talked about plans to use PDRs for distributing newspapers. In the early 1990s, several versions of PDRs were featured in various *Star Trek* episodes and other science fiction movies. The saga of the EO and GO companies heralded the rise of the pen-based computing movement. GO was founded in 1987. Its impressively designed operating system performed well, even on a 286 processor, and its tablet featured a page-sized display. Its usability was hampered by the lack of wireless communications and robust handwriting recognition. Its difficulties, however, involved industry politics as much as technology shortfalls, as chronicled in Jerry Kaplan's book *Start-up* (1995). In 1993, AT&T purchased a majority interest in EO and began selling personal communicators targeted to the mobile business executive. They eventually dropped the project under threat of impending competition from Apple and Microsoft. In the mid–1990s, Sony built a version called SonyBook™ or Sony Data Discman™. It had a small screen and came with very limited titles. I have heard stories about other, later projects of this sort being developed by a wide range of computer companies. It seems that the compromises in technology necessary in the mid–1990s, coupled with undeveloped markets, led to an early demise of many similar projects. Apple built several hundred copies of a device based on a modification of their Duo powerbook computer and used them for internal development but never marketed them. There is an underground Quicktime™ video of an Apple engineer who has been informed that the project was canceled; he tosses one of the devices out an upper-story window. It hits the ground and the case is cracked, but it still works!

By the late 1990s, several almost-PDR devices were appearing commercially, especially from Japanese companies. Fujitsu Personal Systems (a subsidiary of Fujitsu, Ltd.), Epson, and Symbol offered pen tablets and palm-sized computers ranging in weight from about eighteen ounces to about four pounds. The lightest devices tend to have very small screens without backlighting and to lack wireless communications. The four-pound devices have screens ranging in size from 7.2 inches diagonally to 10.4 inches and communicate by using infrared and radio. Screen size matters. Palm-sized and other devices with small screens are useful but are not adequate for prolonged reading or viewing. Other companies, such as Mutoh, offered tethered interactive tablets. A U.S. start-up company, Everybook, proposed to develop a two-screen design that folded and opened like a book with a screen on each side.

Based on or inspired by pen computers, research on portable document readers continued in the mid- and late–1990s in several academic and industrial research laboratories. Digital Equipment Company's Stems Research Center developed a portable reading device called Lectrice as well as reading software (Virtual Book) designed to be easy to use and easy to read on a digital display. At the University of California at Berkeley, the InfoPad project developed a radio-linked portable terminal for wireless access to networks. The Fuji Xerox Palo Alto Laboratory developed the Dynomite electronic note-taking system on a Fujitsu pen computer.

During the writing of this chapter, several start-up companies appeared, offering versions of electronic books or "e-books"—Librius, NuvoMedia Inc., and SoftBook Press. In general, these companies offered systems that allowed downloading of books from on-line bookstores, together with some degree of protection for the downloaded copyrighted material. This generation of products lacks wireless content delivery and seems to focus on analogs of the "book" genre rather than on interactive multimedia models.

At the time of writing, there are still many technical obstacles to creating PDRs. Although incremental improvements in the products now appearing are not likely to yield the perfect PDRs so many dreamers described to me, the devices are getting much closer to the ideal and will continue to advance and fill new applications. When near-ideal PDRs appear in quantity, they will provide radical alternatives for many things and will provoke both opportunities and chaos at the Internet edge. In the following subsections, we consider the technical challenges presently faced by those who would build PDRs.

Flat-Panel Displays

The flat-panel display is the "face" of the PDR. It creates the visual appearance of the book, video, photograph, or whatever else is shown. The requirements for a satisfactory display depend on the specific application. In general, the visual requirements for matching the quality of images on paper are the most stringent—far more demanding than those for ordinary broadcast television. What would an ideal PDR display be like?

It would have very high image resolution, which is an especially crucial factor in designing a display for text reading. In the late 1990s, high-quality laser printers could print twelve-hundred dots per inch or more. In most cases, the dots are accurately positioned but irregularly shaped blobs. Experiments with displays suggest that a resolution of three hundred pixels per inch, with the pixels accurately shaped as squares, would satisfy the visual acuity of the eye and appear as sharp as laser printing. Typical desktop CRT displays and most flat-panel displays, which have resolutions of about seventy-two pixels per inch, fall far short of this level.

Many displays currently in use have substantial power requirements, either for holding and refreshing the data being displayed or for backlighting. Display technologies capable of holding data without the need to refresh them are called *bi-stable;* those illuminated by ambient light are termed *reflective displays.*

Improvements in all these factors are part of the mix for developing displays for use in PDRs. There are presently several competing technologies for building flat-panel displays.

- *Liquid crystal displays* (LCDs), the most widely used technology for portable computing, are not yet ideal, and there are several competing technologies with strong potential.
- Like liquid crystal displays, *plasma display panels* (PDPs) have been in development for several years. They operate on the same basic principle as a fluorescent lamp or neon tube. A PDP consists of an array of hair-thin fluorescent tubes set in a substrate. They can be many square feet in area, very bright, and produce excellent color. PDPs with several million pixels measuring over forty inches on the diagonal were developed in the mid–1990s and are now available from several manufacturers (including Fujitsu, Panasonic, and Mitsubishi). PDPs are expected to become widely available in about the year 2000 for wall-hung television viewing.

Although the price for early models in the late 1990s was very high (on the order of $8,000), it is expected to drop drastically as the consumer market develops. Because of limitations in resolution and high power requirements, PDPs are not expected to be useful for portable devices. They are expected to replace CRTs in applications requiring displays ranging in size from twenty to a hundred inches.

• *Field-emission displays* (FEDs) are a newer display technology with much promise but several technical hurdles to overcome. Like CRTs, FEDs have a screen with a phosphor coating. Unlike CRTs, they work by focusing arrays of microscopic cathode-tip emitters at each screen pixel. The electrons light up the phosphor coatings on the screen to form an image. FEDs are expected to be low cost, lightweight, and rugged and to have low power requirements. Research and development of FEDs is taking place primarily in Europe and the United States. They are being developed in small sizes at several companies, including PixTech, Futaba, and Motorola. Their power requirements are about half that of comparably sized LCDs: a ten-inch display consumes approximately two watts. Furthermore, FEDs do not require backlighting. Remaining technical hurdles include high reflection of ambient light and a gas build-up known as *display flashover,* which can sometimes damage a pixel by creating an electrical discharge. FEDs nonetheless are a potential threat to LCDs for portable applications and have the potential to upset the entire LCD industry.

• *Organic electroluminescent displays* (ELDs) also have the potential to surpass LCDs in performance. The light-emission efficiency of organic EL displays is about the same as that of LCDs, but while the LCD requires constant backlighting, organic ELDs generate their own light. Power consumption is also lower than for LCDs, because only the necessary pixels are lit. Historically, the main obstacle to practical EL displays has been their limited service life, especially for materials that emit red. The red emission layer is difficult to maintain because the longer the wavelength the higher the drive voltage and the shorter the service life. Except for the red layer, materials that last five thousand hours or more have been developed, and Pioneer Electronic began volume production of small-sized organic ELDs in the late 1990s.

A number of other display technologies—including several kinds of electrophoretic displays, micromechanical displays, electrochromic displays, and field-effect dyes—are in various stages of research and development. All of them have significant problems that must be overcome before they can be applied to PDRs. Breakthroughs in any of these technologies could create major changes in the display business.

At present, however, LCDs are the preferred technology for portable computers. LCDs continue to improve, and they have the advantages of market share, mass production, and revenue from ongoing sales. Overcoming these advantages will be a formidable task for any competing technology. What, then, is the state of the art for LCDs themselves?

Although LCDs' image resolution still falls somewhat short of paper's, it is better than that of any other available technology. A flat-panel display capable of rendering an edge-to-edge image of an 8.5" × 11" piece of paper at 300 pixels per inch would require about 8.5 million pixels. Table 2.1 compares this resolution with some of the flat-panel displays commercially available in 1997. The closest in overall pixel count is a black-and white-display by dpiX, which has 7 million pixels. The best flat-panel color displays have 5 million pixels.

Production economics for flat-panel displays are a major factor in the affordabilty and pricing of PDRs. The cost of manufacturing LCDs depends in large measure on the yield of the manufacturing process, during which the entire large area of the semiconductor or amorphous silicon must emerge from all the steps without a single defect such as could be caused by a dust particle anywhere on the plate. Of the displays listed in the table, the three highest-resolution displays are in very limited production for military and special purpose applications. In 1997 the NEC fourteen-inch display was available for $1800 each in quantities of a thousand a month. Smaller ten-inch displays with lower resolutions (such as VGA or SVGA) are generally available at about a third that cost. Profit margins on

Table 2.1
Resolution of commercially available LCDs in 1999

Display	Pixel × pixels	Total pixels
Standard VGA	640 × 480	300 thousand
NEXC14" XGA display	1024 × 768	800 thousand
NEC SXGA display	1280 × 1024	1.3 million
dpiX Color display	1460 × 1190	1.7 million
IBM "Roentgen" display	2560 × 2048	5.2 million
dpiX B&W display	2940 × 2380	7 million
8.5" × 10.5" paper	3150 × 2550	8.5 million

LCDs are relatively narrow, and prices are trending downward, depending on resolution and size. Generally, in the past fifteen years displays have dropped in price by between 10 and 20 percent a year. It is likely that over time higher-resolution displays will be incorporated into PDRs and laptop computers as prices drop.

The power consumption of an LCD depends on its size, with a large portion of it being used for backlighting. For example, the power requirement for a twelve- or thirteen-inch display is on the order of three to four watts. Depending on the brightness required, the backlight can raise the power load of the display by an additional ten to twelve watts. Efficient reflective LCDs that do not require backlighting are under development, although currently, in the late 1990s, the viewing angle on these devices is limited. For example, a five-inch color display from Sharp capable of showing 512 colors has a power requirement of 0.1 watts. Reflective LCD displays have other shortcomings in image quality: the whites are gray and the contrast between white and black is much lower than it is in most printed documents. These difficulties make the displays difficult to read except under carefully controlled lighting.

In summary, LCDs remain the leading contender in the search for PDR displays. At the time of writing, all types of display have relatively poor resolution, compared with paper, and require substantial power. In the absence of a breakthrough—such as a reflective or bi-stable device—LCDs will probably continue to dominate the market.

Communication Technologies

Wireless communication was not included in the first portable computing products. With the advent of cellular telephones and related technologies in the mid–1990s, however, wireless communication came to be seen as a necessary option for portable devices. Although PDRs without such capability would be useful for some applications, adding it greatly expands the possibilities. Two-way communication takes a PDR beyond the dream of a portable, paperlike interface and into the realm of a portable networked appliance.

Bandwidth

What level of communication does a PDR require? A key parameter of communication is bandwidth—the rate at which data can be transmitted and received. Although some enthusiasts of PDRs insist that PDRs can use caching techniques and intermittent connections to minimize bandwidth requirements, others believe that consumers will use as much bandwidth as they can get. This difference in view about bandwidth derives from different views of how PDRs will be used. Some applications, such as sending short e-mail messages, require only modest bandwidth. Others work well enough with only intermittent access to high bandwidth or use low bandwidth to trickle large digital objects into local storage for later viewing. One application with high bandwidth requirements is downloading video, that is, using a PDR to watch digital television.

High digital bandwidth, measured in bits per second, is usually achieved by transmission over a broad band of frequencies, measured in hertz or cycles per second. This approach to achieving high-capacity digital transmission is referred to as *broadband.* It is also possible to use other techniques—such as sending data through multiple narrower frequency channels or switching a base-carrier frequency on and off.

How much bandwidth is needed for satisfactory transmission of digital video? Because costs tend to rise with increasing bandwidth, many tradeoffs are possible. We can reduce bandwidth requirements by various compromises in picture quality, such as reducing the number of pixels in an image, decreasing the rate at which an image is refreshed, or limiting the color gamut of the image. There has also been active work on reducing bandwidth requirements by using video compression. A lower bound for bandwidth requirements in digital transport of video is the version used in PicTel™ systems for dial-up video conferencing. This approach uses Integrated Services Digital Network (ISDN) for a bandwidth of 128 kilobits per second (kbps). However, the visual artifacts and blurring that occur when anything in the visual field moves make this low bandwidth unacceptable for entertainment. The upper bound for bandwidth requirements for digital video is 19 megabits per second (Mbps). This data rate comes from the Federal Communications Commission's Advisory Committee on Advanced Television Service, whose recommended standards for high-definition digital television were adopted by the FCC in auctioning off the frequency spectrum. This figure is considered an upper limit because high-definition television was originally designed for presentation on large screens, rather than on the small screens appropriate for portable devices. An upper bound of bandwidth for standard television is 1.5 Mbps, the data rate provided by the Asymmetric Digital Subscriber Line (ADSL), fast digital telephone lines originally designed to deliver movies to the home.

Another key parameter of wireless communication is range. Some portable devices are designed for relatively high bandwidth connectivity but only within a given work site. It is also possible to design devices that would use high bandwidth within a work site and lower bandwidth over a larger area.

The standards and technology for providing portable communications are now evolving rapidly. In the United States, portions of the broadcast spectrum are being carved up and auctioned off by the FCC for a variety of purposes. At the same time, technological development is being fueled by new demand and, in some cases, by an industrial shift as technology originally developed for military applications is being deployed for civilian and commercial purposes.

Mobile Bandwidth Today

With the advent of cellular telephones, it became possible for determined road warriors with laptops to connect their portable computers to portable telephones to create fairly portable network appliances. In the late 1990s, this became easier and cheaper when wireless services like Metrocom's Ricochet network provided small transceivers that can be attached to a laptop computer and a cellular radio network. In a way analogous to a radio network for voice communication with cellular telephones, the Ricochet network deploys a set of radio transceivers, or microcells, in clusters from a half mile to two miles apart. The transceivers are typically located on buildings, utility poles, and street lights. A Ricochet radio modem sends data packets to the nearest microcell, which then hands them off across the network to a wired access point with land connections to the Net. In 1997, the Ricochet network was available throughout the San Francisco Bay Area, the Seattle metropolitan area, Washington, D.C., many airports, and on many university campuses.

Although the over-the-air data rate for a Ricochet modem is 100 kbps, actual performance is slowed by network traffic and the store-and-forward process of moving the digital packets through the microcells to the Net. The closer a computer is to a repeater radio or wired access point, the higher the bandwidth. Typical communication bandwidth is similar to that of a telephone modem, with speeds ranging from 14.4 to 28.8 kbps. Performance depends on where a person is in the network, the applications being used, and specific hardware.

The Ricochet network is one example of the use of a relatively low bandwidth connection over a wide area. In the late 1990s, Fujitsu offered a PDR-like device providing higher-speed radio communication over limited areas. The device provides data rates up to 1.6 Mbps on fifteen independent channels in the 2.4-GHz band. Fujitsu's system is based on a wireless land area network (LAN) system by Proxim, which connects portable devices to each other or to servers on a wired network over distances up to one thousand feet. Because a work site often has sources of radio interference from equipment as well as reflections from physical objects, the wireless system uses a frequency-hopping, spread-spectrum approach to improve system robustness and bandwidth. Wireless spread-spectrum systems for local area networks are also available from other companies, such as Lucent Technologies.

We can summarize the wireless digital communication technology available for a PDR in 1997 as offering two choices: low-bandwidth radio communication suitable for e-mail over reasonably large areas, or higher-bandwidth communication within a work site or a very limited area.

Wireless but Not Mobile

Wired connections—*land lines*—currently provide most high-bandwidth communication between Internet sites. Table 2.2 summarizes the data rates available or projected for several kinds of wired and fiber-optic connections.

Integrated Services Digital Network is a dial-up digital modem service that became available in the late 1990s in many regions. It uses one or two channels on a telephone, expanding to the second channel as the data rate requires. Asymmetric Digital Subscriber Line (ADSL) service became available on a trial basis in the late 1990s from several companies in the United States, including Ameritech, NYNEX, Northern Telecom, GTE Telephone Operations, and Pacific Bell. ADSL is designed to transport data and

Table 2.2
Data-transmission of communications technologies proposed or available in 1999

System	Bandwidth
ISDN	128 kbps
MVL	768 kbps unidirectional; 384 kbps bidirectionally
Universal ADSL	1 Mbps
ADSL	1.5 Mbps downstream; 64 kbps upstream
DMT-ADSL	6.144 Mbps downstream; +96 kbps bidirectional
T1 line	1.54 Mbps
T3 line	45 Mbps
OC-3 (e.g., ATM on copper)	155 Mbps
OC-12 (e.g., ATM on fiberoptic)	622 Mbps
Gigabit Ethernet	1 Gbps
OC-48	2.5 Gbps
OC-96	5 Gbps

standard video services as well as regular telephone service over a single copper line. It delivers a unidirectional high-speed data or video channel while simultaneously providing telephone service and a lower-speed return signal. It was originally developed to deliver movies to the home on demand. DMT-ADSL (Discrete Multitone) can theoretically improve transmission reach by dynamically switching to frequencies with less interference. In the late 1990s, the prospects for widespread adoption of ADSL went through several cycles; numerous competing proposals called for standards based on systems with different speeds upstream and downstream, different requirements for home installation, and different degrees of technological readiness. Confusion in the marketplace resulted as companies deployed incompatible versions in different parts of the country while consortia tried to agree on standards. In the late 1990s, several computer and communication companies formed a working group to promote a "universal" mass market version of ADSL that would be easier to install in homes and would offer an "always on" connection, though at a much lower bandwidth than DMT-ADSL.

Internet hosts typically connect to the Net over T1 and T3 lines. These lines offer, respectively, data rates of 1.54 Mbps and 45 Mbps. The data rates with The letters OC in the name of a carrier stand for *optical carrier.* The base rate, for OC–1, is 51.84 Mbps; each higher level operates at a speed divisible by the base rate. Thus OC–3 runs at 155.52 Mbps. As advances in transmitting over copper wires were made, these speed designations began to be used for these lines as well. In the late 1990s, data rates above OC–12, earlier obtainable only in experimental systems, became commercially available in some locations.

Asynchronous Transfer Mode technology (ATM) was designed to improve bandwidth over existing local and wide-area networks while guaranteeing bandwidth for real-time traffic. An ATM network consists of a set of ATM switches interconnected by point-to-point links. ATM networks are fundamentally *connection-oriented,* meaning that a p oint-to-point virtual circuit needs to be set up across the ATM network prior to any data transfer.

In the 1990s, making a broadband connection for most sites has meant laying fiber-optic cables. Recognizing the expense and practical difficulties of doing so in cities, several companies have started experimenting with wireless delivery of high bandwidth to fixed locations. Local Multipoint Distribution Service (LMDS) is an emerging broadband service that operates in the 28–31-GHz frequency range. It is intended for delivery of voice, video, and data to business and residential end-users. The digital bandwidth provided by LMDS depends on the frequency bandwidth and degree of sharing of the channel. CellularVision offers 500-kbps service in parts of Manhattan; it is aimed at cable television, data, and telephone use, but that is not an upper limit. Some services proposed would deliver 1.5 Gbps downstream and 200 Mbps upstream per household. LMDS relies on a cellular architecture that supports new capabilities in fixed wireless telecommunications. In late 1997, as the FCC put 1.3 GHz of spectrum up for licensing in 492 trading areas nationwide, the future of LMDS was chaotic and poised for rapid expansion. A key technical challenge for growth of a mass market in this area is development of low-cost transceivers.

WinStar is an example of a company that is providing this technology. Its wireless technology uses the 38-GHz spectrum to deliver OC–3 (155 Mbps) data rates on a 100-MHz channel. In a typical application, the antennas are mounted on rooftops or windows in an unobstructed line-of-sight configuration. The antennas interconnect two customer locations or link a location to a WinStar fiber access node. All communications in this frequency range require line-of-sight to ensure high reliability. Signals at this part of the radio spectrum propagate like fat laser beams and may be blocked by foliage. In residential applications, therefore, reliability requires antennas to be mounted above the tree line.

LMDS is an improved version of an earlier microwave technology known as Multipoint Distribution Service (MDS) or wireless cable. MDS is a low-powered broadcasting system operating in the UHF bands—2.076 to 2.111 GHz and 2.3 to 2.4 GHz. It was designed for low-powered analog services using transmitter powers of 200 watts within a service area of thirty kilometers. Multi-channel Multipoint Distribution Service (MMDS) is an MDS system that broadcasts on more than one channel to potential customers. The main problem with both MDS and MMDS is that the signal tends to break up in bad weather. LMDS addresses this problem by using repeaters. LMDS also relies on new generations of faster semiconductors capable of operating in the 30-GHz region. Speaking of the reliability of LMDS systems at a public forum before the FCC, Stagg Newman of Bellcore remarked that "enthusiasm for LMDS is inversely proportional to the number of leaves you see when you look out the window" (1997). In October 1996, Pacific Telesis Group agreed to buy wireless transmission capacity from WinStar. Initially concerned about the reliability of a wireless connection in rain and other weather conditions, Pacific Bell tested the service and found its reliability to be on a par with its wired services.

Broadband wireless connections are also being offered via satellites. These systems can be used in regions that lack an available fiber-optics network. General Instrument Digital Wireless offers satellite connections with 17 Mbps data rates. An example of a satellite-based technology for homes and small businesses is DirecPC, a 400-kbps Internet connection at the 12-GHz frequency. DirecPC is offered internationally by several companies, including Hughes, Olivetti, and Digital Satellite Source.

Toward Mobile Broadband

In the late 1990s communication technology was co-evolving rapidly along with public policy and the allotment of frequency spectra. None of the technologies discussed so far can quite offer broadband communication for a mobile application like a PDR. While suitable for e-mail over a wide area, low bandwidth communication such as Ricochet is on the slow side for Net surfing and is inadequate for most video applications. Wireless LAN technology offers much better bandwidth but is limited to a work site. LMDS technology has plenty of bandwidth, but does not work for mobile applications.

At the end of the 1990s a new technology entered the mix. *Phased-array antennas,* originally developed for the military to enhance radar sensitivity, have the potential to greatly improve bandwidth at the work site and also to provide higher bandwidth to mobile units.

Several systems using this technology for mobile broadband communication are currently at the research stage. For example, the System For Advanced Mobile Broadband Application (SAMBA) is a joint European project to develop a broadband cellular-radio communication system. It is intended to provide ATM connections and other services up to 34 Mbps. A key idea in SAMBA is the use of phased-array antennas as adaptive, directed antennas. Directed antennas focus their transmitted signal in a narrow region rather than over a circular region, thus increasing the range for a given power and reducing interference with other signals. The ability to focus the signal also means that the same frequencies can be reused by a base station to communicate simultaneously with multiple receivers located in different directions. Direction division also increases the capacity for sharing channels using Frequency Division Multiple Access (FDMA). The latter shares the same frequency with multiple communicators and with Time Division Multiple Access (TDMA), which timeshares portions of a channel at a given frequency among several communicators.

In the late 1990s, broadband mobile communication was represented by experimental prototypes in research laboratories, but it will be some time before it can be commercially deployed. The advantages of phased-array antennas are understood, algorithms for tracking and locating roaming computers are known, and experimental access protocols are in use. Although there appears to be a tight trade-off between bandwidth and the number of mobile users and delay, the bandwidth of 34 Mbps in the SAMBA project is more than adequate for most PDR applications. However, there are still technological obstacles to applying directed antennas. To increase their communication range, it will be necessary to provide both the base and the mobile unit with directional antennas. Mobile tracking will also be needed to aim the beams with adequate precision. These technological challenges will need to be surmounted before this approach can be put into general use.

Impulse radio, a radical approach to high bandwidth wireless communication, was reported in 1999 by Time Domain Corporation. For communication, impulse radio uses long sequences of pulses called "pulse trains." Each pulse is a short nearly Gaussian monocycle ("pulse") with a tightly controlled pulse-to-pulse time interval. According to the company reports, Time Domain has built prototypes with variable data rate at 39 and 156 kbps, communicating over ten miles using only 250 microwatts. Impulse radio is predicted to be suitable for site communications at 160–400 megabits per second, as well as for communications to miniature cell phones. It requires relatively simple electronics and antennas but uses highly precise timing circuits. In 1999 the promise of impulse radio was reported in the technology news but the timing of its practical application for mobile communication was uncertain owing to a combination of development issues and patent litigation.

In summary, the economic and pragmatic issues related to a communications technology for providing wireless broadband either at a work site or over large areas have yet to be addressed, although the prospects for mobile broadband are improving. As communications technology develops, it seems likely that users of PDRs will have multiple choices, with increasing bandwidth available as they shift from wide-ranging mobile use, to use within a site, and to wired connections. Finally, even for mobile use, there may be distinctions among available channels. For example, a PDR may be able to receive a constant high-bandwidth public channel for digital mass media and a slower or intermittent two-way digital connection for personal communications.

Processor Technology

Watching trends in the early microelectronics industry, Gordon Moore predicted in 1965 that the number of transistors on computer chips would double every eighteen months. For over thirty-five years Moore's prediction has been accurate and has led to the widespread view that such incremental progress in chip design is almost inevitable. The driving force behind Moore's Law has been the industry's success in decreasing the feature size of integrated circuits, that is, computer chips. As chip features get smaller, the computers built out of those chips get faster—at the rate of roughly 44 percent per year.

Why do chips get faster when feature sizes get smaller? Although the particulars of chip design are complex, there are some simple ways to explain why this is generally so. Electrons travel at about half the speed of light. When chips get smaller, the electrons that race between registers during computations arrive sooner because they have shorter distances to travel. This observation would predict a roughly linear speedup as feature sizes shrink. But the speedup is actually more dramatic than that. A more accurate way to understand the phenomenon is to remember that state and communication on chips is carried out by charging wires and gates. Wires and gates have *capacitance,* the ability to hold a charge. The capacitance of gates and wires declines with conductor size, and, because they are two-dimensional areas, they scale with the square of the feature size. Thus, small feature size yields smaller devices that can be charged more quickly.

Chip feature sizes have decreased in size because of improvements in lithography, the technology use to inscribe patterns on the silicon that chips are made from. Shorter wavelengths of light can be used to write smaller features on a chip, just as sharper pencils can write thinner lines.

In 1997 the ultraviolet (UV) light chip makers used to image features on state-of-the-art chips had a wavelength of 248 nanometers, which enabled them to imprint features of 0.25 micrometers. By the year 2001, industry leaders expect to produce features of 0.18 micrometers, using UV with a wavelength of 193 nanometers. With a few more improvements, they hope to produce features of 0.13 micrometers by 2004. Past that point, by 2010 or 2017, different observers predict that different obstacles to incremental progress will come into play. There is, however, general agreement that new approaches, such as using X-rays or so-called *extreme ultraviolet,* will be needed to inscribe chip features.

Even in the next few years, fully exploiting smaller feature size for greater chip speed will run up against other challenges. Chips made with the mainstream production processes of early 1997 had 0.35-micrometer features and speeds of 233 MHz. At a feature size of 0.18, a 1000-MHz (or 1-GHz) chip would consume 40 watts of power (in the absence of other changes). To reduce power consumption, industry leaders will have to set a goal of reducing voltage from 3.3 to half a volt. Progress on this front was announced in 1997, when IBM revealed it had developed the ability to use copper, instead of the usual aluminum, for metal connections on chips. The change enables them to use a voltage of 1.8 volts with 0.20-micrometer features. In 1998 IBM announced it had developed silicon-on-insulator (SOI) technology. This advance involves effective use of a thin layer of insulating material that reduces the amount of electrical charge a transistor uses to store information. This technology reduces the effect of interference between signals on a chip and the effect of background radiation—giving a chip designer an option to lower the voltage, raise the operating frequency, or shrink the circuits. Use of SOI is expected to increase the speed of chips by as much as 35 percent. These advances exemplify the suite of technologies the industry must incorporate as it moves toward the next generation of computers. In early 1998, several companies announced laboratory success with *gigahertz* chips, including one made by IBM that consumes only 6.3 watts, even without using copper or SOI.

The availability of capital for manufacturing advanced chips is another challenge, as each new generation requires additional capital. A plant for making 0.25-micrometer chips costs between $2 and $2.5 billion to construct. Manufacturing plants for 0.18-micrometer chips are expected to cost between $3 and $4 billion.

What are the implications of these trends in processor and chip technology for PDRs? Following out the curves to 2002 suggests that PDRs would run at 1.2 billion instructions per second and have a gigabyte of RAM and twenty gigabytes of storage. In general, the incremental progress on chips is all good news for the feasibility of PDRs. PDRs can make good use of the increased speed the next generations of chips will provide. Faster hardware for signal processing would be very useful in implementing phased-array antennas for mobile communication. Faster hardware for encryption could be very useful in designing trusted systems for ensuring electronic commerce and privacy on PDRs. Faster hardware for bulk compression and decompression of data could help reduce data rates for mobile communication. Integration of the processor and memory on a single chip could speed up the general level of processing for interactive systems.

The technology-development curves for processors and chips are encouraging and may help to simplify PDR design by compensating for challenges in other areas.

Battery Technology

Today's users of laptop computers have an intimate but unwanted relationship with batteries. In the early 1990s, Radio Shack's TRS 200™ computer achieved almost twenty hours of use on AA batteries available everywhere. It achieved this battery life by using a characters-only display with no backlighting. In the late 1990s, however, typical laptop users are lucky to get four or five hours on a charge. Thus, today's users must keep track of the state of charge of their batteries and the opportunities for recharging them. To compensate for the shortage of readily available power outlets on airplanes and public places, they often carry a charger plus two or three extra batteries, thus effectively doubling the weight of the computer.

As a colleague of mine recently remarked, there is no equivalent of Moore's Law in battery technology. Batteries do not improve every year by incremental and predictable advances in materials design but only when there are breakthroughs and when new technologies are introduced. The timing of breakthroughs, though, is not predictable.

There are two key measures of battery technology: the total energy capacity of a battery and the energy it can store per a given weight or volume. The latter measure is called the *gravimetric energy density* (GED). An alkaline D-cell can store about 65,000 joules. In an ideal case, this means that we could draw from it about one watt of power per second for 65,000 seconds.

In figuring weight considerations for PDRs or other portable devices, it is useful to consider the GEDs for rechargeable battery technologies that were available in the late 1990s (see table 2.3).

Table 2.3
Gravimetric energy density (GED) of rechargeable batteries, 1998

Technology	Watt-hours/kilogram	Watt-hours/liter
Nickel cadmium	40	80
Nickel metal hydride	60	110
Lithium-ion	90	130
Zinc-air	146	181
Zinc-air (button cells)	340	1050

Zinc-air batteries, which came on the market in the late 1990s, require significant free airflow over the battery. They can power a 20-watt portable device for over eight hours and weigh 3.4 pounds. The U.S. Advanced Battery Consortium and others are funding development of lithium-ion polymer batteries targeted for use in electric cars. These batteries have projected capacities of 200 watt-hours per kilogram.

The battery situation for PDRs is similar to that for laptop computers, since the underlying technologies are so similar. Both kinds of devices need power for the processors, for local storage, and for the display. Furthermore, the power needed for digital communications will put substantial additional demands on battery life. In the late 1990s, a lightweight laptop weighed two to four pounds. Adding wireless communication such as Metricom's Ricochet wireless modem (including the battery) adds an additional thirteen ounces. Under normal use, the battery for a laptop lasts from two to four hours and consumes ten to twenty watts per hour, with the lion's share going to the display. The most advanced model of the wireless modem comes with a nickel cadmium battery that lasts three hours in continuous use or about twelve hours in typical intermittent use.

Two trends beyond battery chemistry affect power usage in portable devices. The first is use of so-called smart batteries, which include an integrated circuit that monitors various parameters such as voltage, charge, and temperature and provides precise control of battery charging and output. This increases battery life by about 25 percent and enables the charge to be precisely controlled. The second trend is designing the electronics of the device to optimize battery use. Ricochet's wireless modem, for example, goes into a "doze" mode to conserve power when it is out of range of the cellular transceivers and when there is no data to be sent. Some laptop systems allow a user to make trade-offs in performance—such as adjusting clock speed, display brightness, and disk performance—in order to prolong battery life. For portable devices used in a contained environment, another option is to provide ubiquitous docking stations where the devices can be easily and conveniently recharged.

One of the implicit and wonderful properties of paper media is that documents can be read in ambient light without requiring power. Although readers could not expect to download a novel at the beach and read it all day without changing batteries, they could probably do so if they had an extra set of batteries and recharged them overnight. In the absence of breakthroughs in technology—either to increase the power capacity of batteries or to decrease the power requirements of displays and other components—PDRs in the near future will not be as trouble-free as their paper counterparts.

Summary of Technology Trends

We have seen in the foregoing sections that the technology for building an ideal PDR does not exist in the late 1990s but is getting closer.

The LCD displays now available for PDRs have relatively high power requirements. Although displays whose apparent resolution approaches that of printing on paper are very expensive, they are decreasing in price at the rate of 10 to 20 percent per year. LCD technology has been improving incrementally, and several competing technologies are waiting in the wings.

The communications technologies currently available for PDRs are not capable of broadband data rates over a large geographic areas. They can support low bandwidth (suitable for e-mail) over a large area and higher bandwidths at a work site. Technology for directed phased-array antennas is in the works and could radically increase available bandwidth and decrease power requirements both over large areas and at work sites.

Processor technology continues to follow Moore's Law, yielding speed increases of about 44 percent per year. In the next few years, increases in speed may help to overcome other design challenges, such as the need for faster processing in signal processing, compression, and encryption.

Battery technology does not follow any trend of predictable incremental improvement. With currently available laptop systems, the trade-off between weight and power capacity produces a portable life of about four hours between recharges. Contributions to the useful life of batteries for PDRs may come fairly soon from the use of low-power electronics and reflective displays. But a radical breakthrough in either display or battery technology will be needed before manufacturers can produce the ideal PDR weighing less than a pound.

Being Careful What You Wish For

There are many folk tales in which a character is granted three wishes. He or she spends the first wish accidentally, on something foolish, and the second deliberately, on something seemingly desirable that has unintended and disastrous consequences. The third wish, of course, is spent on recovery—trying to return things to the way they were before the wishes.

People have been dreaming about and wishing for PDR-like devices for many years. It is hard to predict from the technological trends exactly when the technology for PDRs will be widespread and cheaply affordable. Nevertheless, the possibility of their widespread deployment comes closer every year. Children born in the next few years will be used to life with PDRs, just as many children born in the 1980s never experienced life without microwave ovens, home video, or computer games.

How will PDRs change our lives? To find answers to this question, we can ask ourselves two related questions: What could a PDR do? What would we do with them?

PDRs and related devices are more than technological achievements. We started our discussion of PDRs by comparing them to paper and considering ways to make documents both digital and portable. Making documents digital and portable changes a lot of things, and adding the potential for broadband communications changes a lot more. PDRs will change our personal and work lives and, inevitably, cause chaos during the period of change. They will create major discontinuities with what came before them. Such discontinuities are coming; it is just a question of timing and staging. Consider the following scenarios about some of the things we may get when we wish for PDRs.

Digital News

There is an old joke about a group of travelers who get lost in the country. Unable to find their way to a particular small town, they drive up to a farmer standing by a junction and ask him for directions. First he points off in one direction and starts giving directions. But then he pauses, and says "Nope. That won't work—the bridge is out." Then he points in another direction and starts talking again. But he realizes another problem. Finally, he says. "Sorry. You can't get there from here."

The humor of the joke lies in its absurdity. Although we may have to go the long way around, we can always get there from here. In cyberspace, we want the same thing. We always want to be able to get there from here. Increasingly, devices for cruising the Net are also able, one way or another, to receive television and radio programs. Digital services like Pointcast also make it possible to receive and read multiple digital newspapers.

What is the future of the newspaper in this situation? Given a portable device like a PDR, we can imagine it as an extension of what Pointcast provides, delivered to our PDR electronically through wireless broadcast to read at our convenience. It would be updated during the day as the news develops. It would be interactive, and its stories would be linked to related stories. It would let us retrieve yesterday's story and perhaps even go to an encyclopedia or reference collection to find background information on an interesting story. The stories would have pictures. If we clicked on the picture, we would get a short video and watch the president get off a plane, see the crowds at a demonstration, and so on. There would probably be advertising. If we clicked on an advertisement for a restaurant, we could make reservations for dinner and parking.

What about the future of the television news program? Given a portable device like a PDR, we could watch it at our convenience. The news would be delivered electronically through wireless broadcast and updated throughout the day as it develops. It too would be interactive, linking news stories with related stories. And so on. The future newspaper and the future television news program might become the same thing. Once-separate industries would first begin to compete and then to merge with each other.

Thus PDR technology and its connections with the Internet will simultaneously blur distinctions between existing media and create new media. New writing styles and genres will evolve to suit the new media. Newspapers in the 1890s were written in a style very different from what we find in today's newspapers. Back then they had much less competition for delivering international news, and stories were often blunt, biased, and highly politicized. Worldwide communications now put greater checks on what is written. It is much harder to maintain credibility with a distorted point of view when other, easily accessible voices, pictures, and opinion leaders are reporting on the same events.

From one perspective, the blurring of boundaries between digital newspapers and digital news programs simply reflects the convergence of the media and the electronic industries. That convergence will be accelerated by the widespread availability of a lightweight portable device for receiving digital works. Thus the technologies of displays, communication, processors, batteries, and electronic commerce will work together to crystallize the media convergence. That is part of what we will get when we wish for PDRs.

Digital Schoolwork

According to a report in the November 18, 1997, *San Jose Mercury-News,* state school officials in Texas were considering substituting digital textbooks on laptops for paper textbooks. The argument was partly economic. The officials projected that the state would spend $1.8 billion for textbooks over six years for 3.7 million students. Textbooks go out of date quickly, but laptops, they reasoned, would be easy to update and could also connect students to resources on the Internet. The school board hoped to improve education and reduce costs simultaneously by a radical move into the information age. They had the right idea, even if their timing was wrong.

What is it that laptops lack to make this dream realizable? Or, stated in a different way, What is the pushback from the edge? There are many things. Laptops are fragile. The batteries don't last long enough. The screens are too small for easy reading. Those available in 1997 were not ideal PDRs because, in part, of the technological challenges yet to be overcome.

Ultimately, however, there are some wonderful possibilities for using PDRs in education. Given their communications capability, it would be easy to update textbooks periodically. Exercises and homework assignments could be sent out over the Net. Textbooks could be linked to other, on-line reference sources. Evening communication by e-mail would be possible with the teacher, volunteer assistants, or other students. Collaborative projects could be enhanced. These advantages are already part of student life for many college students, especially in engineering schools.

A switch to digital delivery has major implications for publishers. Although there are potential cost savings for digital delivery, most of the costs related to textbooks are incurred during their development, rather than in their typesetting, printing, binding, and shipping processes. If digital texts were upgraded to include multimedia features, development costs would probably be even greater than they are for paper texts.

There are other implications for the educational publishing industry. Currently, textbook publishers develop and provide teachers and students with supplemental materials such as worksheets and test materials. With digital delivery, it would be possible to tailor interactive exercises to particular students and classes and, in some cases, to grade exams automatically. Book publishers and educational suppliers may want to charge for such supplemental features, for example, by putting digital property rights on documents delivered to PDRs that are trusted systems.

Some of the preparation and distribution of these supplementary materials could also be left to teachers. Many teachers already spend long hours devising such materials but face obstacles in obtaining such supplies as paper for reproducing and distributing them to their students. Although preparing digital materials requires teachers to become skilled users of computers, it also offers them some advantages. With digital delivery, there are no consumable supplies to run out of.

In short, PDRs could radically change the delivery of educational materials and the experience of education. This is part of what we will get when we wish for PDRs.

Not Forgetting Things

In the 1990s, many office workers carried heavy briefcases full of paperwork back and forth to work, organizing the papers they need to be sure they end up in their briefcases when they leave the office. Inevitably, they forget things. Only when they are at home or on the plane do they discover that something crucial—such as the pre-reading for tomorrow's meeting—is still on the desk at the office.

As more documents become digital, forgetting to pack documents could become a thing of the past. Everything would be available on the Web. With a communicating PDR, you could get at your documents no matter where you are. You couldn't leave them behind because they're out there in cyberspace ready to be retrieved. That is part of what we would get when we wish for PDRs.

Active Reading and Sensemaking

Paper has many convenient affordances that support reading. When we read our copy of a report, we can add marginalia and perhaps use a highlighting pen to mark important sections. When we look at a copy of an article someone else has marked up, what he or she has written in the margins may be as interesting as the original words. When we read books, it is handy to stick in bookmarks or other things to mark our place. Practiced readers have skills for reading and mastering large amounts of material. They read at different speeds—skimming some sections and concentrating on others, marking important information. When analysts and writers read magazines, they often arm themselves with scissors for clipping out materials, or they photocopy sections to file away and refer to when writing their own reports, articles, or newsletters.

The early 1990s saw an interesting shift in the typical use of personal computers. The World Wide Web was beginning to make information sources widely available on-line. Prior to that time, personal computers were used mostly for word processing, accounting, playing games, and communicating via e-mail. As the Web took hold, there was a shift in information access. People began to surf the Net and to use it as an information source. The amount of information available via a personal computer expanded by a factor of thousands. For the first time, the volume of what we received from our personal computers exceeded what we typed into them.

However, the usefulness of the personal computer for reading does not necessarily correspond to its usefulness for retrieving information. For many people the first step toward reading information retrieved from the Net is printing it out. Computer screens, mice, and keyboards lack paper's crucial affordances for reading.

In the late 1990s, Bill Schilit, G. Golovchinsky, and Morgan Price (1998) of the FX Palo Alto Laboratory built a prototype PDR called XLibris. XLibris was built out of a commercial high-resolution pen tablet and a paperlike user interface that supported key affordances for active reading similar to those provided by paper documents. For example, the device allows the reader to make free-form digital "ink" annotations while reading. A key insight in this project was that computational support for such annotation could do more than enhance reading on a PDR; it could also tie reading into information search processes. XLibris thus provides affordances to help readers organize and find information while retaining many of the advantages of paper. Marked-up sections can be collected together, and marked-up phrases can be matched against documents on the Net or in an information repository—automatically converting highlighted phrases into queries and retrieving potentially relevant documents for later reading. The concept of active reading recognizes that reading is not an isolated process but part of a larger process of retrieval, organizing, and writing—what chapter 5 describes as sensemaking. Integrated systems supporting reading and sensemaking for knowledge workers is part of what we get when we wish for PDRs.

Cookies and Milk

Most broadcast mass media today are paid for by advertising. Printed mass media, such as magazines and newspapers, are also largely paid for by advertisements although other printed media, such as books, are not. Except in a general way, mass market advertisers do not know who is reading a particular article or watching a particular television program. They create advertisements intended to appeal to a statistical profile of the watching audience.

With digital delivery platforms, advertising could easily change. Advertisers could know a lot more about the person watching or reading a digital work and deliver different advertisements to different people. Moving television and magazines to PDRs could lead to changes some people might see as effective, directed advertising and others might see as intrusions.

How would advertising on the Net work? Imagine that a document distributor creates a so-called sponsored browser or reading program that runs on digital systems like PDRs. The sponsored browser is analogous to Netscape's Navigator™ program or Microsoft's Internet Explorer™ in that it displays and supports interaction with digital works encoded in a particular format. What's different about this browser is that it is associated with an economic model of advertising. Owners of rights in the digital works available on the sponsored browser receive payments when someone reads their work, and. advertisers pay to have their messages sent to people whose age, income, address, and so on coincide with the advertiser's customer profile. When a sponsored browser is accessed by a PDR, the browser collects certain information about the person and passes it along to a central processing repository. The browser then correlates this information with other data gathered to target advertising more effectively. A sponsored browser's advertising arrangement would differ from mass marketing in two interesting ways. First, advertisers do not necessarily endorse or care about what materials the user is browsing, although they may collect data about it. The focus of interest is the data about the consumer (age, income, etc.), which the browser company uses in deciding which ads to display. In general, advertisers are willing to pay more when their messages reach people who are more likely to patronize their products.

Here are some examples of how sponsored browsing might work.

- A student is doing homework on a PDR and using a sponsored browser to read some material. As it gets late in the afternoon, the browser notices that the student has been working for several hours. A cheery little jingle plays. "Hungry? What could taste better than cookies and milk?" Late at night, a related jingle might appear. "There's nothing like Coke and pizza to keep a night owl going. Thirty-minute delivery guaranteed. Click here for a large cheese pizza."
- An office worker is putting in some extra hours completing a report. It's been an expensive month and her credit card balance has gotten a little high. A message appears in a window on the display: "Credit card bills got you down? Like to consolidate your bills in one easy, monthly payment? (Click here.)"
- A parent is helping a teenage son with homework. While they are retrieving some on-line reference sources, an advertisement for cigarettes or some other items the parent does not approve of appears on the screen. The parent is puzzled about why the advertiser has targeted the son with these ads.

These examples raise issues about customer privacy and computers, which are the topics of chapter 8. Moreover, it seems likely that not all publishers will choose to make their works available through sponsored browsers. Some readers will respond negatively to the stream of advertisements and prefer to pay browsers for access to information. If PDRs become very popular for all kinds of reading and viewing, advertisers will undoubtedly yield to the temptation to support sponsored browsing. That too is part of what we might get when we wish for PDRs.

Millions of Eyes

PDRs, like other digital devices on the Net, will support two-way communication. In the 1990s, the price of video cameras for personal computers dropped drastically. It is reasonable to expect that such devices will also be incorporated into PDRs and similar handheld devices.

One of the interesting features of networked digital devices is that they probably will not need vast amounts of local storage. With adequate bandwidth, they could store pictures, sounds, and videos directly on the Net. When potentially millions of people are carrying around devices that can easily record information and post it on the Net, there will be a very good chance that whatever they are doing at a given moment will be posted to the Net. Such widespread deployment of information has, again, many implications that bear on personal privacy (as considered in chapter 8). That, too, is part of what we may get when we wish for PDRs.

Reflections

At the beginning of this century Edison believed that the major use of the phonograph would be for business dictation. Ultimately, Berliner showed that its most important application would be for entertainment. Even then the recording business did not stand still in the face of emerging technologies. In the 1930s, the advent of the "wireless," or radio, affected the industry, but not in the way people expected. Instead of making it unnecessary for people to buy records, radio broadcasting promoted record buying. As the businesses co-evolved, radio promotion, record sales, and advertising found complementary, and mutually supportive niches in the music business. In 1928, the Radio Corporation of America (RCA) purchased the Victor Talking Machine Company, forming RCA Victor. In 1938, Columbia Records became part of the Columbia Broadcasting System (CBS). Manufacturers began to produce entertainment consoles that included both radios and phonographs.

Wireless communication, like radio broadcasting, brings mobile connectivity. When media become wireless they become untethered. Cellular telephones enable mobile workers to connect with home or office from any place at any time. They change the implications of the advertising line, "reach out and touch someone," an act no longer limited to the place where the telephone plugs in. The combination of radio and PDR displays is equally dramatic. The cellular PDR will do for information and knowledge workers what the cell phone did for voice and mobile workers. It has the potential to accelerate the convergence already happening in the media industries. It has the potential to change many things in our lives.

Jon Kabat-Zinn wrote a book on meditation practice in which he describes what a person learning to meditate faces. Meditators begin to notice things about their minds. Thoughts arise, and habits of thought become more apparent. They find that they can distract themselves from seeing themselves, but they can't really get away from their thoughts: "Guess what? When it comes right down to it, wherever you go, there you are" (Kabat-Zinn 1994).

As PDRs become ubiquitous, we can be connected together a lot more of the time than we are now. Wherever *we* go, there *we* are. Without proper balance, though, there can be a collision between our collective identity and our personal identities, and with our individual needs for quiet reflection and privacy. In this chapter we pointed to this potential dark side of PDR development by considering the possibilities that sponsored browsers will target advertising at an intrusive level and that greatly increased public Net postings and recording of our affairs may trespass on our private lives. In chapter 8 we focus in some detail on several of the social, technological, and legal issues regarding privacy the Net raises.

But networked connectivity has a bright side too. The bright side includes the convenience of being able to get the information we need no matter where we are. We won't need to worry about forgetting documents anymore. PDRs could also support new, powerful tools for reading and sensemaking. Among the other themes considered in this chapter are the recognition that the Net is not just for business, and that PDRs and the Net will become more powerful when combined with digital broadcasting . PDRs could change our use of information in education, allowing students to collaborate more easily, even on short projects. It is interesting to note that these themes were present in the earliest incarnations of the Net. Many of the first applications of the Internet were educational. In fact, one of the earliest implementations of the packet-based communication protocols used on the Internet today was Norman Abrahamson's use of radio signals for the ALOHAnet at the University of Hawaii in the late 1970s.

We have seen that the combination of the Net, broadcasting, and PDRs could grant us some freedom from the laws of physical space. Cyberspace connectivity clearly changes the balance. The question is, What balance points shall we choose?

3

The Digital Wallet and the Copyright Box: The Coming Arms Race in Trusted Systems

Tell me, people of Orphalese, what have you in these houses? And what is it you guard with fastened doors? . . . But you, children of space, you restless in rest, you shall not be trapped nor tamed. Your house shall be not an anchor but a mast. . . . For that which is boundless in you abides in the mansion of the sky.

Kahlil Gibran, *The Prophet*

The term *trusted system* came originally from military terminology. It refers to computer systems that provide access to secret information for national and military purposes. In the last few years, its meaning has been broadened to include systems that protect and govern the use of digital objects and information for commercial purposes.

In chapter 2 I considered the development of personal document readers (PDRs), devices that many believe will become the delivery vehicle for digital published works in a multimedia blend. However, releasing valuable works to a digital medium creates a risk for publishers and authors, whose works could be copied and distributed without compensation.

The technological response to this risk is to use trusted systems, which protect digital works using a set of rules describing fees, terms, and conditions of use. These rules, written in a machine-interpretable digital-rights language, are designed to ensure against unsanctioned access and copying and to produce accurate accounting and reporting data for billing. Such trusted systems are the "copyright boxes" in the title of this chapter.

Creative works are not, of course, the only digital objects that need secure storage, accounting, and machine-governed rules of use. The idea of digital cash, or tokens, that can be used as money in cyberspace has caught the public imagination. Handheld trusted systems that permit the exchange of such tokens are the "digital wallets" of the title. Like physical coins and bills, and unlike checks, digital cash can be exchanged anonymously and, like copyrighted works, according to specific rules. Just as unauthorized copying of a published work amounts to copyright infringement, unauthorized copying of digital tokens amounts to counterfeiting money. Other rules governing cash—for example, those forbidding the anonymous transfer of large sums of money into or out of countries—could cause users of digital cash to run afoul of import regulations and laws about money laundering.

Mondex, one of the companies that offers technology for commerce in digital cash, uses trusted systems based on *smartcards,* plastic cards the size of credit cards with built-in computer chips. These cards have been used in Europe for phonecards and other purposes for several years but, through the late 1990s, had a limited presence in the U.S. market. Mondex's digital wallet is used to read the amount of cash on a card and to transfer money between cards.

In a conversation about digital cash, John Reed of Citicorp quipped to me, "How do you know when Mondex has a bug?" Playing the straight man, I asked: "How do you know when Mondex has a bug?" "When M1 rises," he answered, naming the Federal index that measures the amount of cash in circulation. Such gallows humor about smartcards is not at all far-fetched. It illustrates an underlying fear about digital cash and trusted systems. In essence, the workings of digital technology are largely invisible. We may not realize that a system is broken or compromised until after the damage has been done.

In 1998 a phonecard-piracy scam came to light in Germany, where phonecards designed by Siemens for Deutsche Telekom pay phones are based on smartcards. Ordinarily, once a phonecard's balance reaches zero it is thrown away or given to collectors. A group of "pirates" from the Netherlands found a way to bypass the security of the EEPROM chip used on the cards without leaving physical evidence of tampering and to recharge the cards. They bought thousands of spent cards from collectors, recharged them, and resold them to tobacco shops and retail outlets across Germany. The losses were assessed at about $34 million. This was not the first attack on the cards. The European digital wallet arms race is producing successive generations of cards that are supposed to be more resistant to tampering.

All kinds of programs—not just digital cash and creative works—can benefit from secure storage and guarantees that they work properly and have not been tampered with. We want our computers to have trustworthy programs that are under our control. We want our computers to bring us information about the world but also to be discreet about revealing private information about us to others. As more of our everyday world comes under the control of software and is networked, the issue of computer trustworthiness will extend beyond our desktops and into other parts of our lives. As described in chapter 10, computers may eventually manage, not only our businesses but also our vehicles, homes, and even, through wearable computers, our bodies.

How secure are trusted systems? How secure do they need to be? How do we keep people from circumventing the safeguards of the box, or prevent computer viruses from inflicting malice and mischief on the accounting systems, the protected works, and the system user? How hard is it to break into a trusted system? If the contest between builders of trusted systems and hackers intent on breaking into them is essentially an arms race, is this a race that can be won?

The Coming Arms Race in Trusted Systems

> [This layer is a] complex layout that is interwoven with power and ground [wires] which are in turn connected to logic for the Encryption Key and Security Logic. As a result, any attempt to remove the layer or probe through it will result in the erasure of the security lock and/or the loss of encryption key bits.
>
> Manual for the Dallas Semiconductor DS5002FP, a security microprocessor

> [We] designed and demonstrated an effective practical attack that has already yielded all the secrets of some DS5002FP based systems used for pay-TV access control. . . . the attack requires only a normal personal computer . . . standard components for less than US $100, and a logic analyzer test clip for around US $200.
>
> Ross Anderson and Markus Kuhn, "Tamper Resistance—A Cautionary Note"

Companies developing trusted systems for protecting copyrighted works include International Business Machines, FileOpen, Folio, InterTrust, NetRights, SoftLock, Xerox, and Wave Systems. Initiatives to develop trusted systems are underway by Intel and Microsoft; companies developing digital wallets include Cybercash, Digicash, and Mondex. Other companies are building systems for using digital cash in on-line shopping.

Designers of trusted systems for military and national security applications assume that the "security threat" will come from a determined, well-funded, malicious, and technically astute adversary. The U.S. Department of Defense *Orange Book* (DOD 1985) discusses the system requirements that must be met by defense contractors building trusted systems for the military. But what are the requirements for trusted systems that handle digital money or copyrighted works? The answer is "it depends."

The Economics of Pirateware

What are the risks that people will develop "pirateware"—hardware and software for circumventing trusted systems? Compared to digital cash and military applications, the threat to trusted systems for digital publishing is sometimes thought to be minimal. But is it? One way to assess the situation is to look at the economic motivations. What are the perceived risks, costs, and benefits (or value) for those who would infringe? What are they for those who would manufacture and sell pirateware? And what are the risks, costs, and value of digital publishing for rights owners? A basic dictum in the design of secure systems says that, to be effective, they must make the costs and risks of pirating much greater than the expected benefits.

As I discuss in chapter 4, the conventional wisdom about paper publishing is that the cost of making a photocopy of a substantial work is high, compared to its purchase price. Publishers believe that what is mainly needed to reduce losses from isolated acts of copying is a way to make it easier for basically honest people to stay honest. For example, if there were a simple and automated way to pay a modest fee to rights owners, such as by inserting a credit card into a copy machine, honest people would pay the royalty without further ado.

This line of thought suggests that the level of security required for protecting copyrighted digital works is similarly quite modest. But this is misleading. The risks and benefits of copying digital works are not really the same as they are for paper works. Without trusted systems, digital technology actually increases the publisher's risk by practically eliminating the infringer's costs of copying and distribution. A digital publisher has no advantage over an infringer when it comes to manufacturing low-cost copies. With a few keystrokes, any computer user can copy a paragraph, an article, a book, or a lifetime of work and mail it electronically to thousands of people. In the absence of trusted systems, many publishers—fearing that digital distribution really means routine and potentially massive copyright infringement—withhold their valuable works from the Net. Because the losses from infringement of digital works are potentially so great, the benefits of such encroachments are also high. According to our basic dictum about security-system design, if the perceived value of the protected goods is high, then the expected cost of defeating the security system must be made even higher.

The Internet Edge for Trusted Systems

The dictum about designing for security does not assure us that we will actually have trusted systems. This is where resistance at the Internet edge for trusted systems comes into play. It is just as possible that we will not have trusted systems and that valuable digital goods (or substantial digital money) will not become available on the Internet. For any particular technological proposal, the pushback can be that the required security measures—which may include such things as special hardware, special software, or impractical changes to computers already in use—are simply too expensive.

The back-and-forth probing at the edge between the forces for creating trusted systems and the forces holding them back is fueled by the perceived economic value of digital publishing and the perceived expenses of adequate security. Each journey to the Internet edge is an attempt to find a way to serve some of the potential market.

The possible outcomes of the journey include prospects for digital publishing and digital money, prospects for piracy, and, potentially, for changes in the legal status of practices that undermine copyright. Technology consultant Matthew Miller (1997) illustrates this point with a perspective on the evolution of technology for protecting satellite-television transmissions with descrambler boxes. In the late 1980s, anyone who wanted to watch satellite television could set up a large dish antenna, hook up a descrambler box, and pay monthly fees. However, the technology of the descrambler box was so simple that it was widely duplicated and sold in the underground and hobbyist markets. What were the risks and benefits to the pirate? Because television programming caters to a broad market, and because the same descrambler box would work for anybody with a satellite dish, the black market in descrambling boxes had broad appeal. Furthermore, there were legal ambiguities about whether signals broadcast in the air were in the public domain anyway. When the satellite broadcasters approached Congress asking for legislation to prohibit the manufacture and sale of the rogue descrambler boxes, they got little support. Legislators argued that broadcasters had done too little to protect their signals. Enforcement would be expensive, and legal relief could not adequately compensate for technological weakness. It made no sense to protect satellite transmission, even by a thin legal veil, until its trusted system technology attained a reasonable level of security.

In the years since this example played out, several different approaches to secure commercial trusted systems have been developed. No trusted system is perfectly secure, and some security arrangements are more costly than others. As it was in the satellite example, the legal status of systems for defeating copy protection is murky. Attitudes toward copyright and pirateware are still a matter of debate on the international scene. Even in countries with strict laws for protecting copyright, enforcement is uncertain.

One size does not fit all for trusted systems. The market for trusted systems will probably be stratified, with the least-expensive systems being used for the least-expensive works and the most-expensive trusted systems used for the most-expensive works. This prediction is based on several observations. The first is that technology for high security costs more than technology for low security. The second is that, with low-priced works, a certain amount of leakage may be offset by the broader market served by inexpensive or even free trusted systems. The security levels of such systems are not, however, appropriate for distributing digital objects when there is high incentive to steal them.

Technological Foundations of Trust

In the popular press, the use of encryption technology is often equated with high security in computer systems. This is a misleading association. To use the analogy of a door, consider the flap of a tent, the front door of the average house, and the combination door on a bank vault. Arguably tents, houses, and banks may all contain things of great value. In a nearsighted theoretical sense, it might seem desirable to put a security door on a tent to protect campers from wild animals. However, a solid metal door such as is used on a bank vault would destroy the portability of a tent without improving its safety, since a persistent adversary could easily come in through the canvas wall. The sobering truth about security is that there are many potential ways to defeat it, especially when it is not a primary design concern from the beginning.

Trust is based on two things: responsibility and integrity. When digital works are stored and used on trusted systems, the systems are responsible for accurately ensuring that they are used in accordance with the rules expressed in the terms and conditions. When digital cash is stored in digital wallets and spent on goods and services, trusted systems are responsible for accurately following fiducial rules for handling cash.

The integrity of trusted systems depends on three technological foundations: physical integrity, communications integrity, and behavioral integrity. Physical integrity refers to the capability of a trusted system to resist physical tampering. Communications integrity refers to its ability to detect any misinformation or lies it receives in its digital communications with other systems. Behavioral integrity refers to the persistent ability of trusted systems to enforce terms and conditions and to resist unauthorized modifications to their programming.

Physical Integrity

The possibility that a pirate will penetrate the hardware and thereby gain access to information stored in a repository is one kind of threat. Sensitive information in trusted systems includes not only protected works but also billing logs, encryption keys, digital cash tokens, and personal and financial information.

Different repositories can have different levels of physical integrity. A repository that can be compromised with a screwdriver would have a low level of physical integrity. A somewhat higher level of physical integrity would be a system with built-in sensors that enable it to detect a threat and to erase sensitive data. A still-higher level of physical integrity would cause a system to self-destruct when it detects a threat, perhaps setting off alarms and telephoning for help.

Computer peripheral component interface (PCI) cards are devices roughly the size of a small paperback book that are used by plugging them into a computer. Using trusted systems allows PCI cards to hold certain central and sensitive data in financial services—for example, passwords, keys for authorizations, and possibly secret algorithms. To preclude unauthorized electronic probing of information stored on the PCI card, designers cover it in a material that has several layers of nichrome wire. To read the signals in the card's sensitive circuits, an attacker must first penetrate the cover material. Drilling a hole to gain access to the circuits is likely to break one of the wires, which would be sensed by the circuits and trigger a signal to erase sensitive data. This design feature is a first layer of defense against a physical attack.

The arms race in designing secure circuits is like a spy story or a master-level strategy game, with systems within systems and feints within feints. One line of attack takes advantage of the physical properties of computer memories, which are susceptible to low temperatures. If memory circuits are chilled to low enough temperatures, they are unable to change state; they cannot, for example, respond to a signal to erase their contents. Knowing this, attackers can first chill the card, then drill into it and, even, remove its components, confident that the system will be unable to erase its secret information. Before the card warms up, they can disconnect all the defense mechanisms, enabling them to read the data at leisure. A defense against this attack is to put thermal sensors on the card to signal an attack when the temperature drops. There are many other possible measures and countermeasures for designing trusted systems. The interplay between "attacks" and "countermeasures" makes the term *arms race* an appropriate metaphor for the design of trusted systems.

As trusted system defenses are elaborated, handling and shipping them can also become more difficult. Elaine Palmer of IBM tells a story about some secure PCI cards built by IBM for bank computers. A shipment of cards was bound for a bank in Moscow. The bank had already closed when the truck carrying the shipment arrived late in the afternoon. The truck had to be parked outside overnight in the cold Moscow winter. As the temperature fell, the cards' thermal sensors signaled "thermal attack." The operational information was erased before the cards could be installed.

Inexpensive smartcards, generally costing under ten dollars, are being used in an increasing number of systems, ranging from pay television to digital wallets. Because overcoming the tamper resistance of these applications results in substantial returns for the infringer, there have already been several cycles in the trusted system arms race for smartcards.

The typical smartcard has a single plastic-encapsulated chip containing an eight-bit microprocessor with memory and serial input and output. Key data stored in an erasable programmable read-only memory (EPROM), whose contents can be changed by using a twelve-volt signal. As smartcards lack batteries, they cannot use active defenses involving sensors, clocks, and preprogrammed responses to threats.

Anderson and Kuhn (1996) describe a wide range of attacks on the physical integrity of smartcards and other so-called tamper-proof equipment. The early smartcards used for pay-TV systems received their reprogramming signals along a programming voltage contact on the card. Subscribers who initially had their cards enabled for all channels could cover the contact with tape and then cancel their paid subscriptions, leaving the vendor unable to cancel their service.

Physical attacks can be divided into noninvasive and invasive approaches. Noninvasive approaches tend to exploit the responses of smartcards to unusual voltages and temperatures. In some processors, for example, repeatedly raising the supply voltage when a smartcard writes to a security causes the lock to release without erasing the memory it was protecting. Conversely, in another processor, a brief voltage drop sometimes releases the security lock without erasing secret data. "Glitch attacks" use transient signals to interfere with the operation of particular instructions, such as instructions for outputting data or checking passwords.

Current smartcards have almost no defense against direct access to the silicon circuits, that is, against invasive attacks. Some cards use capacitive sensors or optical sensors to detect the continued presence of a covering layer. These sensors tend to be easy to detect and avoid.

Attackers can cut away the plastic with a sharp knife or hand lathe and then remove the chip. Or they can remove the plastic resin over a chip by applying nitric acid alternated with acetone washes. As nitric acid is used to clean chip surfaces during manufacture, it affects neither the silicon (or silicon oxide or nitride) nor any gold used on the chip. The information in the EPROM remains intact and is available for reading. This sort of attack is sometimes used by so-called class I attackers—amateur pay-TV hackers, students, and others with limited technical resources. More sophisticated attacks can be carried out by pirates with access to focused-ion-beam workstations and infrared lasers.

At present, untested claims for tamper-resistant smartcards and other security processors should be taken with a grain of salt. For many systems the first "hostile reviews" appear only after they are put on the market. In the current escalating and open-ended technological arms race, state-of-the-art engineering practice for trusted systems will change rapidly over time. Prudent designs will employ appropriate techniques to ensure that the cost to the pirate will be much greater than the expected benefit. Systems that can be defeated by a simple attack to one component should be avoided.

Communications Integrity

Not all attacks on trusted systems require physical contact. Communication attacks are attacks made over the wire, when a nontrusted system tricks a trusted system into giving up or compromising digital goods, private information, or digital money. To carry out their functions, trusted systems must communicate with people and other systems. In digital-wallet applications, these communications include the transfer and validation of digital tokens representing money. In digital publishing, they include the transmission of digital works and billing data.

In general, a trusted system "views" the world through its communication channel. To use an analogy, imagine a trader locked in a room who has to obtain information and conduct all his or her business by telephone. Like a trusted system for buying and selling works or for transferring money, the trader can have certain passwords and keys. The world of this telephone trader (or trusted system) is rendered more complex by pirates, who may make fake telephone calls, masquerade as other parties, or listen in to calls.

Trusted systems, like our trader, need reliable ways to certify the bonafides of the party on the other end of a communication line, to verify that the messages exchanged are genuine and unchanged, and to keep the data communicated on the line secret from eavesdroppers. Meeting these goals requires communications integrity. In general, the foundations of communications integrity—secure and robust communications across insecure lines—lie in encryption technology.

When trusted systems connect with each other, they go through a registration process by which they identify themselves to each other and establish their bonafides. Once they are connected, they put each other through a series of tests—a *challenge-response protocol*—intended to weed out impostors and to protect the entrusted works. When registration succeeds, they establish a trusted session using encrypted communications.

Behavioral Integrity

How do we know that trusted systems will operate properly—even if they have not been physically compromised and can communicate securely and prove their identity? For the most part, the behavioral integrity of computers is determined by their programming. There are no fail-safe ways of writing general programs that guarantee correct behavior. There are, however, a number of ways of reducing the risks of error in sensitive operations: for example, by designing systems modularly—so that all critical operations take place in a small part of a program—or by defining system functionality in layers—allowing certain operations to be carried out only under restricted conditions. In addition, designers can build into other parts of the system checks and balances that log sensitive operations and even prevent them from being carried out. Finally, there are ways of writing programs that check other programs before they are run, essentially proving that certain properties are intact under all possible operating conditions.

All of these approaches have blindspots and shortcomings. A rogue programmer can violate modularity rules with hidden code, or violate security layers with trap doors. Checks and balances can be compromised by hiding transactions. Rogue compilers can insert statements into a program that a proof checker (or even a careful programmer) will never see. Furthermore, when the specifications for ensuring a program's correctness become as complex as the program itself, it becomes very difficult to have high confidence in the specifications themselves. In the end, our trust in a program is based on the capabilities, methods, and reputation of the organizations that write and certify programs.

In an ideal computer, the operations carried out by programs could be securely isolated. For example, a screen saver program could not interrupt an electronic-commerce application to steal data or fake an authorized transaction. Similarly, a word-processing program could not, under the influence of a virus, modify a system file to provide a trapdoor for tampering with financial records. An applet used to drive an animation on a web page could not tunnel into files elsewhere in the computer, remove information, and alter their contents. However, no such protections are built into the operating systems presently in wide use today.

In the absence of trusted operating systems with such secure boundaries between programs, any program can in principle compromise any other program. In this situation, the only guarantee of behavioral integrity is the certification of all programs loaded onto the computer and the warranty that no program can be altered after it has been certified. However, the brute force needed to carry out this guarantee stalls when it confronts the inertia of the vast installed base of uncertified operating systems and popular applications.

Thus, in theory the foundations for trust in a trusted system are its physical integrity, communications integrity, and behavioral integrity. But the practical reality is that existing standard platforms are poorly suited for trusted systems. In the next section, we consider some of the design tensions and methods for building a generic trusted player under these circumstances.

Case Study: Anatomy and Operation of a Trusted Player

Trusted systems can be built into various applications. For example, we could build trusted systems into boomboxes for the metered playing of music. The personal document readers described in chapter 2 could be designed as trusted readers. Trusted printers that receive and spool digital works securely and put watermarks on pages can be built to carry information about the identity and authorized use of the work and about the printing event. Reception units for direct-broadcast satellites (DBS) are also trusted systems.

One of the most ubiquitous kinds of trusted systems is the trusted player, a system for rendering or displaying a digital work. It is typically implemented as a combination of hardware and software on a personal computer. Figure 3.1 illustrates some of the software components of a generic trusted player. In this section, we consider the anatomy and operation of a generic trusted player and describe two transactions—purchasing a digital

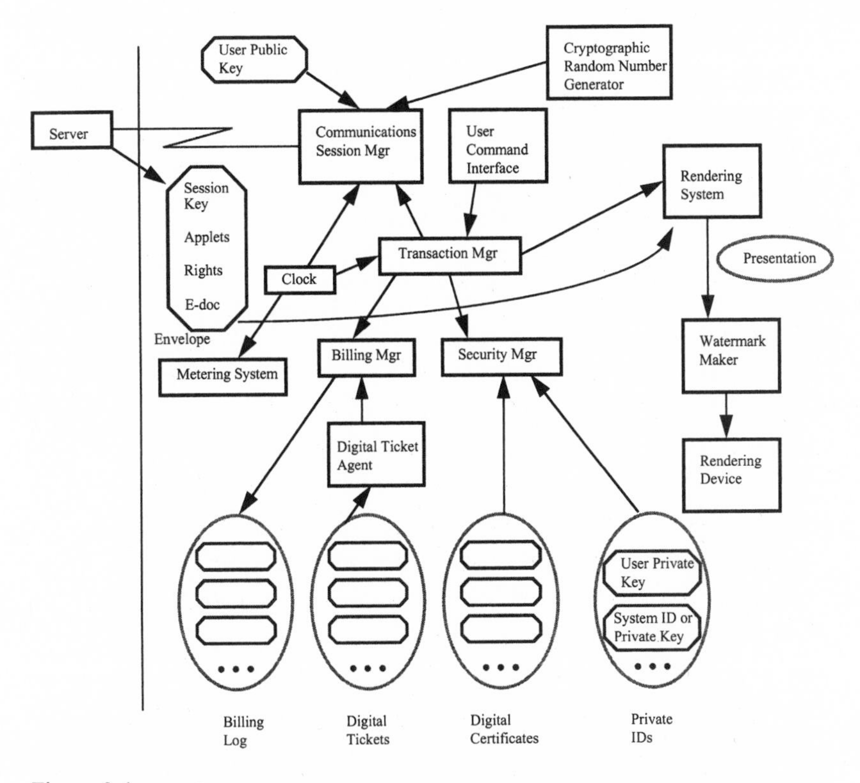

Figure 3.1
A Generic Trusted Player. This system is used to purchase copies of digital works, to store them securely for later use, and to render them on a display.

work and playing it—to illustrate its operation. We then consider some failure modes, possible threats, and means of defense.

No operation on a digital work can be done until the user logs on. The logging-on process is initiated by a command at the user's command interface, which activates the transaction manager and the security manager. First, the user must establish his or her bonafides, typically by supplying a password. (We defer until later a discussion of the security arrangements for invoking or validating the different modules involved in this transaction. However, it is worth noting that this process activates the user's private key for the session, access to which is a crucial security measure.)

If the user wishes to purchase a digital copy of a work, the player initiates a session with a trusted store that has a copy for sale. This process involves a challenge-response protocol in which the two systems exchange data, test each other's certificates, evaluate each other's security classes, and establish cryptographic keys for the session.

Since the concept of a challenge-response protocol is well known in the art, our description of it will be brief. Each trusted system sends the other a digital certificate, which is digitally signed by a well-known repository; the certificate confirms the system's identity, public key, and security arrangements. Each system then constructs a *nonce*—a random sequence of digits—and sends it, encrypted in the other's public key, to the other party. Each trusted system decrypts the nonce, using its own private key, and sends it back to the other. The systems also synchronize clocks and check their "hot lists" of rogue systems. If the trusted player fails the nonce test, has too much deviance in its clock, or is on the hot list, the trusted store terminates the transaction.

Additional exchanges establish what works are available for sale and the terms and conditions for each transaction. After identifying a desired work, the user requests a copy from the user command interface. The request specifies the terms and conditions of the sale, including the price, any required certificates, and the security class. If the trusted player fails to satisfy any of the requirements for security, available funds, or valid certificates, the transaction is aborted. If the expiration date on the user's right to copy has expired, the transaction is also aborted. If all conditions are satisfied, the trusted systems begin a copy transaction, in which the store repository transmits an encrypted copy of the work. Typically, the work is block-encrypted, and the encryption key is itself encrypted in the user's public key and included with the work. The trusted player stores the work in encrypted form, and both systems make billing records of the transaction. If the transaction is not completed for any reason, both sides report the interruption in their billing records, and the trusted player deletes its encrypted partial copy.

Playing the work requires the user to exercise a play transaction, which is invoked by another command at the user command interface. In some systems, this merely amounts to pressing the "play" button; on others, the user may select from several different play options. Some works offer free copies—but charge for their play options. A play fee may be a flat fee, or it may be metered according to playing time.

When the user action initiates the play transaction, the transaction manager first checks that the requested right to play has not expired and summons the security manager, which checks that all the certificates required in the terms and conditions are available and valid. It then invokes the billing manager and the rendering system. In this example, we assume that the rendering system and the display are integral to the trusted player. If the rendering system is a separate unit, the trusted system begins another transaction backed by secure communication. If digital watermarking information is specified, the information about the user and purchase is encoded in hidden data (the digital watermark) when the work is delivered.

When the playing process is complete, the transaction manager informs the billing manager, which updates the billing log as appropriate.

Boundaries and Threats

We use threat analysis to determine in advance what can go wrong in a system, whether through malicious interference or equipment failure. The likely attacks on an actual implementation—and so the kind of analysis needed—vary according to the particulars of the implementation. In practice, what matters is that the examination of possible threats be systematic, thorough, and as realistic as possible. The analysis described in this section is intended to be educational and to provide a point of reference for our later discussion of security measures.

What's Worth Stealing?

To begin, it is worthwhile to account for the potential values (positive or negative) of some possible attacks. If a digital work is encrypted, there is little risk it will be copied when it is stored or transmitted; but if it can be stored in an unencrypted form without authorization, the risk is that the publisher or rights owner will lose revenue from unauthorized use or copying. The owner can limit this risk in various nontechnological ways—for example, by carrying insurance against leakage of works. System designers can also build into all trusted players measures that test the bonafides of various works, perhaps enabling them to watch for and report the sources of works identified as rogue copies. Such measures would increase the risks to those trying to defeat security measures (Samuelson 1996a).

Another potential risk is the change or misappropriation of the user's private encryption key. Changing the user's private key would be (at the very least) an inconvenience, as it would deprive the user of service until the situation is straightened out. In the meantime, the misappropriator could make fraudulent purchases, damage the user's credit, or violate the user's privacy. If a user's private keys are taken, the attacker can access all the works on the system and, furthermore, acquire additional works until the theft is detected or a credit limit is reached. An attacker who accesses or changes the billing log could delete or add billing data, resulting in inconvenience and perhaps lost revenues for rights owners. One means of reducing this risk is to arrange for transactions involving multiple systems to be reported to separate financial clearinghouses and then reconciled. In such cases, events not reported by one system would probably be reported by another.

Stolen digital certificates would have very little value to the attacker, because they need to be validated when used. Tampering with digital tickets is another matter. Digital tickets are limited-purpose digital tokens comparable to script or coupons for certain rights. They are prepaid and can be used once. Copying digital tickets is, therefore, very much like stealing or counterfeiting money.

To summarize: the security of trusted systems depends on protecting users' private keys and passwords, system private keys, digital tickets, and the billing log. In addition, attacks can do damage even when no data are stolen. The purpose of an attack might be, for example, to destroy a copy of a work, undermine the reputation of a competitor, create an inconvenience for a user, or obtain commercially useful information in violation of a user's privacy rights.

Attacks

We now examine the operation of some of the subsystems of trusted systems to identify potential points of attack. In general, our approach is to consider each system module involved in a transaction to ask what could happen if the module were compromised.

The first operation in our scenario occurs when the user logs in. One means of attack is capturing the user's password. Knowing the password gives an attacker access to the system and the ability to masquerade as the user, making purchases or using services without authorization. A variation of this attack modifies the user command interface—so as to save the password somewhere in the clear. Another variation compromises the security manager (which hashes the password and compares it against a stored hash value). A third variation changes the stored hashed copy of the user's password to substitute a different password.

The second operation in our scenario involves the copy transaction for purchasing a digital work. One form of attack is to compromise the random-number generator used to create nonces. This could make determining the private system key mathematically easier and would enable the attacker to compromise the communications manager; the latter, in turn, could then invoke the security manager and save a copy of the work in the clear. Another attack would modify the transaction manager so that it aborts the transaction after the work has been received but before the receipt has been confirmed. (This attack is of no use if the protocol is designed to confirm receipt of the complete work before the decryption key is transmitted.) The security manager might then also be impaired and induced to release the system's private key.

The last attack in our scenario would compromise the play transaction for delivering the digital work by modifying the clock. Compromising this component would permit the system to exercise expired rights (for example, a free trial period). This change would prevent the transaction manager from invoking the billing manager or cause the billing manager or metering system to bill inaccurately. The security manager could also be altered to make it omit the checking of certificates or other terms and conditions or to release system keys or keys to the work. The rendering system could be compromised so as to release copies of individual "pages" or "screens" of the work. Or it could be modified to put false watermark data on the presentation or to leave out the watermark entirely.

It is evident from this summary of possible attacks that a trusted system built on a personal computer or workstation has a very long trust boundary. Essentially every module in the system is subject to attack, and an attack compromising any element of the system could cause loss of data, revenue, privacy, or damage to the reputation of a person or organization.

Countermeasures

The difficulty of designing practical trusted systems for digital publishing is inherent in the tension between the need for security and the widespread availability of computing platforms. If trusted systems are widespread but inadequately secure, publishers will not risk releasing their intellectual property on them. If they are demonstrably secure but expensive and rare, publishers will have no incentive to invest in trusted systems, because the market size will be insufficient to earn back the expenses of creating and distributing digital works. Faced with this seeming dilemma, we can come to two conclusions about appropriate courses of action.

The first suggests that designers need to create different classes of trusted systems; works of low value can circulate on low-security systems, and works of high value on systems of substantially greater security. This arrangement would make it crucial to know the difference; that is, to be able to reliably ascertain—by communicating with it—the security level of any trusted system. The second conclusion is that transmission of many works of intermediate value requires personal computers whose security has been augmented by the addition of secure hardware or substantial improvements in the installed base. This approach is more practical than expecting users to buy dedicated trusted systems for accessing secure documents.

Most of the trusted system solutions currently on the market augment the system security of personal computers with software but not hardware. Such systems can defend against casual attacks by uninformed users but not against determined attacks by knowledgeable users with specialized software tools. Nor are they proof against attacks by software viruses unknown to the system. Because this approach fails to provide secure memory, designers have had to limit the functionality of trusted systems in various ways; for example by requiring software-based trusted systems to authorize payments up-front while the system is on-line. As such systems are generally considered inadequately secure to support metered fees, they rely on network-accessed servers to keep track of usage, inventory authorization certificates, and hold prepaid tickets. Finally, these systems tend to be used only for works of relatively low value, because their measures for protecting encryption keys are subject to software attack.

A primary goal of augmenting the hardware of a personal computer is to provide secure storage of valuable data: that is, keys, billing logs, passwords, and digital tickets. The basic idea is to limit access to sensitive data to hardware and software correctly carrying out a particular protocol in a particular context.

Even when a personal computer is augmented with such secure hardware, there is still a long trust boundary to defend. An attack on virtually any module in the trusted system can lead to loss of data, privacy, or funds. One response to this risk is to locate all the modules within a secure co-processor, such as a PCI card, that contains memory, a clock, and disk storage. At the current state of the art in personal computers, this approach would require the security system and the user's own computer to have roughly the same speeds and storage capabilities, making the security co-processor too expensive for most applications.

An important, and cheaper, alternative is to store and execute most trusted system modules on the user's computer but to check them for tampering each time they are executed. This operation is roughly the same as that employed by virus-checking software, except that more powerful methods could be used to ensure the trustworthiness of certified software. General virus-checking programs, which scan files for known viruses, know the identity (instruction patterns) of viruses but not of the programs they are defending. Trusted systems could work the other way around. A better approach, known in the art but not widely used, is to digitally sign and hash all software modules of the trusted system. Thus, when a module is written and installed for use in trusted systems, it would be checked by a certifying body and warranted to faithfully carry out its function. A digital hash function (such as MD5 or a related algorithm) would be used to compute a hash value for the binary code of the program included in the signature. The hash value would be signed by the certifying agency and also by the trusted system itself at the time of installation. So, whereas virus-checking programs can protect arbitrary software (but only against known viruses), the signature-and-hash approach would protect only a particular set of trusted software but would defend it against even previously unknown viruses.

A trusted system with this kind of security system would first check the hash value of any module before executing it. One way to do this would be to arrange for the overall execution of the trusted system to be controlled by a small kernel running on a co-processor on a security card. This kernel would launch security modules by copying them from the disk, checking their hash values, and then starting their execution. In a more powerful approach, the co-processor could exercise considerable control over the execution of the host computer as needed—for example, by running tamper-checking diagnostics on host-system hardware and generally overseeing the execution of all trusted system software.

This operation could be arranged in many variations for defense against different levels of attack. For example, some approaches would suspend operation of the modules in mid-execution to check again whether there has been any tampering during run-time. As always, the goal of such measures would be to raise the bar high enough to discourage determined attackers. Given enough control over the computer and the loading of programs, a trusted system built in this manner would be essentially immune to a software attack, although it would still be subject to sophisticated and more expensive attacks involving combined hardware and software devices.

Before leaving our discussion of augmenting hardware, we should mention two other likely components of a security card for trusted systems: encrypting chips and a clock. The use of specialized encryption chips would allow use of longer encryption keys. Putting clocks on a card remedies a blatant weak spot of most personal computers. The system clock, which on most computers is easy to reset to another date and time, is a feeble barrier to attack. Adding a tamperproof clock to the external card would be a relatively inexpensive defense.

A Cautionary Tale

A fail-safe way to distribute and use digital goods without the bother of special hardware of any kind would be very attractive to publishers and users. In the late 1990s, several companies announced that they had developed such systems, claiming that they would provide protection for owners of intellectual property on the Net.

One such system promoted by Company X was described in a nationally prominent newspaper in 1998. The company's real name is not of interest here, because its technology is very similar to those of other companies, and because the arms race in trusted systems has barely begun. This particular company was founded by people with backgrounds in intelligence work, and their system received favorable comments from several academic computer scientists. However, as suggested earlier, the harder test for any trusted system is the "hostile review," in which determined specialists try to breach the system's security. Very few systems offered for copyright protection are tested in this way prior to commercial release. Under these circumstances, security failure is quite likely to coincide with financial losses.

In the Company X product, a consumer who, for example, wants to view a movie can pay $2 to make a one-time-use digital file or $20 for the right to unlimited viewing. The consumer receives a digital license agreement, selects the one-time-use option, and pays for the movie with his or her credit card. The consumer's computer stores an encrypted copy of the agreement and then receives an encrypted version of the movie. When the user wants to watch the movie, a special computer program (the "player") matches the encrypted movie with the licenses stored in the computer. If a valid license is found, a key unlocks the movie and allows the user to view it. Once the movie has been played, all that is left is a scrambled file—unless the license is updated and another fee is paid.

Safe as such a system may sound, it is vulnerable to several possible attacks.

Copy Attack

After receiving the movie and the license, but before watching it, the user can foil the system by copying the movie and the license to backup storage. To view it a second time, the user first deletes the license and the reencrypted movie from the computer, then retrieves the unaltered license and movie from backup storage, thus restoring the system to the state it was in before it was played—which permits the user to watch the movie a second time. If the user's system has a tamperproof clock, this attack can be thwarted to some extent by using time stamps that limit the use of the movie or certificates to a given time frame. As most computers do not have tamperproof clocks, the user can first reset the system clock to the time at which he or she purchased the movie.

Fake-Player Attack

A more sophisticated attack can be used by a skillful programmer to decompile the player module and make a new version of it without the security-enforcing features. This modified player can be posted on bulletin boards around the Net (until authorities find out and object) or sent around more surreptitiously to people on underground Net mailing lists. This attack creates the risk of a catastrophic system failure for the publisher that could affects all the works using this technology.

Virus Attack

This attack is like the fake-player attack, except that the modifications to the player program are caused by a computer virus turned loose in the network. In this case, the people who watch the movie for free are arguably not guilty of willful infringement. They can say they did not realize that their free use of the movies was caused by an undetected virus.

Liberator Attack

This attack is, again, like the fake-player attack, except that the player is modified to make an unencrypted copy of the movie, which is then circulated on standard video-player applications. This attack can be thwarted to some degree by watermarking the movie so that the identity of the original purchaser can be determined from any copy found in circulation. A hacker defense against such tracing is to use a stolen credit card to buy the movie in the first place. In either case, the publisher is unlikely to recover damages.

At the time of this writing, Company X's product is not in widespread use. Conventional wisdom in the security community is that systems like this will be broken when they are widely used—resulting in a very public "hostile review" and failure.

Reflections

Trusted systems are at a nexus of several forces at the Internet edge. The market opportunity for digital publishing is a powerful force for change. But a countervailing pushback for the status quo is the inertia created by the huge installed base of computers and software designed for neither security nor commerce.

Legal and political forces are moving into the fray in several industries. In chapter 4 I consider the evolution of laws related to Internet commerce, especially with regard to the interplay between copyright and contract law. The evolving political issue of U.S. export policy for cryptographic technology will also influence the development of trusted systems. Because of their original application to military and intelligence communications, cryptographic systems are controlled and classified under import and export regulations as "munitions." Another concern about cryptographic methods is that they could be routinely used with impunity in socially harmful ways—for example, by enabling organized crime to enjoy secure records, secure communications, and invisible money-laundering.

Meanwhile, foreign suppliers of cryptographic technology have begun to use longer key lengths (thereby achieving higher levels of communication security) than their American competitors. The overall effect of this situation on the security of trusted systems is limited, because they are currently constrained more by the lack of certified applications and operating systems than by regulations about the length of encryption keys.

At its core, the drive toward digital commerce and publishing is a shift that lets local businesses increase their global reach by taking a shortcut through cyberspace. However, cyberspace is not simply fast and ubiquitous; it is also largely invisible and intangible. By relying on cyberspace as it exists today, we are moving from local commerce in tangible goods with neighbors that we more or less know and trust to a trade in invisible goods with people we don't know who use computer systems that we need to trust. It is easy to see why this journey to the Internet edge provokes so much pushback and uncertainty.

Political life abounds with issues of boundaries and trust. American currency bears the phrase "In God we trust." Yet President Theodore Roosevelt was widely applauded for his advice to "Walk softly but carry a big stick." When do we give our trust to neighbors, governments, foreigners, computers, even ourselves?

Trust is not simply given; it is built and earned. If we cannot trust computer systems and the invisible and intangible processes that drive them, our only alternative is to increase the visibility and tangibility of those processes. The current chaos at the Internet edge reflects confusion, because we have no assurances of system integrity. To be trustworthy, systems for

Internet commerce and publishing need to visibly demonstrate their integrity and the accountability of the entities they represent.

Computer systems have the potential to provide wonderful visibility and accountability. Just as bank systems provide an audit trail in monthly statements, so trusted systems could provide accounting summaries. Just as political institutions back up the occasional bank failure and respond to claims of errors, so new institutions could stand behind computer systems for commerce. Creating institutions that can certify trusted systems is part of the overall challenge of enabling the information marketplace to grow. Viewing trusted systems in this light makes it clear that there is important social and legal work to be done at the Internet edge.

4

The Bit and the Pendulum: Balancing the Interests of Stakeholders in Digital Publishing

Information doesn't want to be free.
It wants to be paid for.

Member of the audience, Computers, Privacy, and Freedom Conference, March 1997

The drive toward digital publishing reflects our need to be heard. It speaks powerfully to the dream that everyone ought to have instant access to the best ideas, the most creative works, and the most useful information. On a global network publishers can distribute digital works nearly instantaneously at low production costs, giving consumers the convenience of twenty-four-hour automated shopping.

Technology does not, however, exist in a vacuum. Even if all the technological obstacles to trusted systems described in chapter 3 were removed, serious social and legal issues related to digital publishing would remain. At present, then, the potential for digital publishing remains just that—a potential. The market remains nascent because the medium has failed, so far, to balance the interests of important stakeholders. In this chapter, therefore, we consider the dream of digital publishing and the co-evolution of technological, business, and legal innovations needed to balance those interests.

The Pendulum Swings

Computers and the digital medium itself are sometimes seen as the major barriers to digital publishing. When personal computers and desktop publishing first appeared in the early 1980s, many publishers saw digital publishing as too risky. At the time, numerous factors, such as the lack of an installed base of computers and the high costs of production, reinforced publishers' doubts. Another deterrent was fear of widespread unauthorized copying. Realistically concerned about losing control over their intellectual assets, many publishers completely avoided the digital medium. From their perspective, the pendulum representing the balance of power between creators and consumers had swung too far toward consumers.

In the late 1990s, several vendors—including Folio, IBM, InterTrust, Xerox, and Wave Systems—introduced trusted systems for digital publishing. These systems vary in their hardware and software-security arrangements, but they all automatically enforce the specific terms and conditions under which a digital work can be used. For example, rights may be time-limited, or offered to different people and groups (e.g., members of affiliated book clubs) at different fees. Some trusted systems differentiate among such uses as making a digital copy, rendering a work on a screen, printing it on a color printer, or extracting a portion of the work for inclusion in a new work. When asked to perform an operation not licensed by a work's specific terms and conditions, a trusted system refuses to carry it out. So dramatically are trusted systems expected to alter the balance of power between publishers and consumers that some observers have suggested that the pendulum of power is now swinging too far toward publishers.

Copyright and Paper Publishing

Copyright law and user practices derive from several centuries of experience with publishing on paper. In order to promote the creation and distribution of useful works, copyright law grants rights holders—authors and publishers—certain exclusive rights. Authors' exclusive right to control the reproduction and distribution of their works protects their interests and those of publishers, who invest heavily in the development of works as well as in their printing, warehousing, and distribution. During the term of the copyright, the law forbids a second publisher from undermining the market by selling competing and unauthorized copies. Copyright law also addresses the public interest in the free flow of goods and information; for example, through the first-sale doctrine—which ensures buyers' right to dispose of a paper book in any way they wish—and the convention of fair use—which permits the quotation of limited portions of a book for review or scholarly purposes.

Copyright law, by itself, does not prevent unauthorized copying. Where the enforcement of copyright law is imperfect, however, technology limitations and the economics of paper publishing help to check infringements. Although photocopying makes it easy for an individual to make a single copy of a work for personal use, it generally does not facilitate large-scale copying and distribution. The offset presses and other mass-production equipment used by printers and publishers generate books and magazines that are less expensive and higher in quality than photocopied materials. Furthermore, the costs of storing and distributing thousands of paper copies are usually prohibitive for individuals not in the publishing business. Thus publishers of printed materials are protected from infringement by well-funded rogue publishers by legal remedies, and from large-scale photocopying by individuals by quality considerations and the relative economies of scale. Together, these forces create a point of balance between publishers and consumers of paper-based materials.

Copyright and Personal Computers

This balance of power in the paper medium does not, however, directly translate to the digital medium. Even though digital works can be expensive to develop, copying them is essentially free. A consumer with a personal computer and a laser printer can produce a digital copy of a work as inexpensive and as high in quality as the publishers' original. Furthermore, during the PC revolution, community support for copyright has been crucially different for paper and digital media. Whereas copyright in paper media is strong and well established, support for copyright of digital works is weak, or even absent.

Many key institutions have long been active in advocating copyrights for work on paper. In the United States, these include the Library of Congress, the Copyright Clearance Center, and the Association of American Publishers (AAP). Worldwide, although there are national differences in copyright law, an international body (WIPO, the World Intellectual Property Organization) is active in promoting treaties and standards to harmonize national laws. Overall, copyright has worked well enough to support the established book, magazine, and newspaper publishing industries.

As paper publishers began to consider digital publishing, however, they faced resistance from the computer community. Many people in that community believe deeply that computer software and information ought to be free. Although this attitude is contrary to ordinary business practice and market theory, it has grown naturally out of the history and environment of computer science.

Before the rise of the personal computer in the 1980s, digital publishing was mostly the province of academics and scientists, who shared computers and used them to distribute the results of their scientific work. Academic publishing is a special case, in that the sharing of research results is rooted in a philosophy favoring the free exchange of ideas and the open search for knowledge. Academic journals are also unusual in that authors are not paid for their articles, as they are in commercial publishing. Authors, for whom scholarly citation is important for academic reputation and career advancement, often pay to be published through page charges, rather than being paid for their writings.

Software developed for various scientific purposes in the computer-using academic community was created by and freely exchanged among many groups of users. Widely used software enhanced its creator's reputation and standing in the computer community. This was a natural extension for a community in which sharing results and peer review are key to academic advancement. The free exchange of programs that began in the 1960s accelerated the development of computer science, the field directly concerned with the creation and study of the algorithms used to construct software programs.

The wider PC community inherited from the academic community the values supporting free exchange of information. When the PC community first emerged as a hobbyist fringe of academic computer science, these values served it well. Many hobbyists built their own computers and traded programs. However, as the personal computer culture—and industry—matured, the underlying assumptions supporting free sharing no longer applied. Fewer users created their own software or built on each other's programs; instead, they bought software from publishers. Publishers did not rely on academic grants for support; they had to make a living selling their software.

To keep prices down, software publishers amortized the costs of development and production over their user base. They devised several measures—such as rigging computer disks in various ways—to make it difficult to copy programs. However, even legitimate customers found such protection approaches too inconvenient, and they were eventually dropped. Ultimately, large software publishers realized that copying could be good for business. People often learned a software program by using pirated copies, then later bought legitimate copies of newer versions. This process created a market dynamic that helped large publishers dominate the field and marginalize smaller ones.

By the early 1980s software publishers were operating in a legal regime that provided, at best, uncertain safeguards against the copying of digital works. The courts pondered the extent to which copyright ought to protect the structure, sequence, and organization of computer programs (e.g., in *Computer Associates v. Altai* 1992). Certain well-known computer copyright cases—such as Apple's dispute with Microsoft over the Macintosh graphical user interface *(Apple v. Microsoft 1994)*—took many years to resolve. Meanwhile in other electronic venues such as videocassette recording, court challenges suggested that certain kinds of copying were permissible on the grounds of fair use *(Sony v. Universal Studios 1984)*. And, as noted in chapter 3, Congress refused to pass laws to regulate copying of satellite television programs until broadcasters developed adequate anticopying devices.

Comparing computers to leaky bottles, John Perry Barlow (1994) argued that once a digital work is created, it will inevitably be copied.

Copyright and Trusted Systems

Beginning in the 1990s, however, computer scientists realized that computers could become part of the solution to the copyright problem they had reputedly caused. The key was development of trusted systems technology.

The two main ideas behind trusted systems are (1) that the terms and conditions governing the authorized use of a digital work can be expressed in a computer-interpretable language; and (2) that computers and software can be designed to enforce those terms and conditions. Xerox's DPRL (Digital Property Rights Language) is an example of such a rights language.

Digital rights cluster into several categories. Transport rights include the right to copy, transfer, or loan a work; render rights pertain to playing and printing a work; and derivative-work rights govern excerpting portions of a work, limited editing of it, and embedding parts of it in other works. Other rights govern the making and restoring of backup copies. With trusted systems, a publisher can assign these rights to a digital work and stipulate the fees and access conditions governing the exercise of each specific right.

Trusted systems enforce the assigned terms and conditions and allow exchange of the work only with systems that can prove themselves to be trusted systems through challenge-response protocols. Trusted systems thus form a closed network of computers that excludes non-trusted systems and that collectively supports use of digital works under established rules of commerce. When digital works are sent between trusted systems, the works are encrypted. When they are rendered by being printed on paper, displayed on monitors, or played on speakers, the system can embed in the signal machine-readable watermark data that make it easier to trace the source of any unauthorized copies.

In general, the higher the security of a trusted system, the higher its cost. High-security trusted systems can detect physical tampering, set off alarms, and erase secret key information. Intermediate-security trusted systems have more modest physical, encryption, and programmatic defenses. Using challenge-response protocols, all trusted systems can recognize other trusted systems and determine their security levels. This enables publishers to specify for each work the security level of the trusted systems allowed to receive it. A sensitive industry report, for example, might require an expensive and secure corporate trusted system with advanced security features. On the other hand, a widely distributed digital newspaper subsidized by advertisements might require only a modest level of security easily attained by home computers.

Trusted Systems and the Balance of Interests

There are many stakeholders in digital publishing. Besides the federal government, U.S. copyright law focuses on two parties or categories of people: rights holders (that is, the authors and publishers who hold the copyrights) and the public. Even in paper-based publishing, however, there are multiple intermediaries, including wholesalers, bookstores, used book stores, and libraries. Trusted systems, which delegate enforcement and control to computers, introduce other third parties: trusted system vendors, financial clearinghouses, and national governments. Thus the use of trusted systems complicates the balance of interests by introducing new stakeholders.

Further, the use of trusted systems to enforce terms and conditions provides a much finer grain of control than copyright law, and it moves the legal basis of protection toward that of contracts and licenses. This more precise degree of control distinguishes among usage rights for copying, loaning, printing, displaying, backing up a work, and so on. It also provides for the identification of specific users, rendering devices, and usage fees. In addition, trusted systems bring about a finer grain of control by enabling rights holders to monitor transactions for the usage of works.

In the following sections we consider the sometimes competing interests of stakeholders and the technological and institutional implications of those interests in digital publishing that uses trusted systems. In particular, we contrast copyright-based and contract-based protection of digital works and examine how issues such as fair use, liability, and national borders are likely to play out in a trusted systems regime.

Copyright Law

As excellent summaries of U.S. copyright law are available elsewhere, we will not recapitulate them here. However, understanding certain key aspects of copyright law will clarify how stakeholder interests will influence the design of trusted systems.

Like other forms of legal protection for intellectual property, including patent and trademark law, copyright encourages the creation of intellectual property by granting certain exclusive rights to its creators. Article 1, section 8 of the U.S. Constitution authorizes Congress to create legislation "to promote the Progress of Science and useful Arts, by securing for limited Times to Authors . . . the exclusive Right to their respective Writings." Over the years, Congress has enacted a series of copyright laws, beginning with the Copyright Act of May 31, 1790. The most recent major overhauls of the copyright law are the Copyright Act of 1909 and the Copyright Act of 1976.

Under copyright law, the federal government grants a copyright owner certain exclusive rights in his or her work. These include the right to reproduce copies of the work, to prepare derivative works based on the work, and to distribute copies to the public—by sale or other transfer of ownership or by rental, lease, or lending. For certain kinds of works, such as literary, musical, and dramatic works, exclusive rights to publicly perform and display the work are granted as well.

These rights last for relatively long periods of time. For works created after January 1, 1978, copyright expires fifty years after the death of the author or composer or (in the case of works made for hire) the earlier of seventy-five years from the date of publication or a hundred years from the date of creation. (Legislation passed in Congress in 1999 extends the copyright term to seventy years after the author's death or, for works made for hire, the earlier of ninety-five years from publication or a hundred and twenty years from the date of creation.) Although the term of copyright is long, it is finite; once it is over, the copyright owner's exclusive rights end. Thereafter, the work falls into the public domain, where anyone may publish it freely.

Copyright law attempts to strike a balance among the competing interests of various stakeholders, especially those of rights holders and the public. It addresses this balance by limiting the duration of exclusive rights. It also limits the scope of the legal protection afforded. The Supreme Court has held, for example, that copyright protection does not extend to the "sweat of the brow" invested by a work's creator but only to the author's clearly original contributions. Thus an alphabetically arranged white pages telephone directory may lack sufficient original content to be protected by copyright, even if compiling it required considerable effort (*Feist v. Rural Telephone 1991;* Samuelson 1996b). Further, copyright protection extends to the expression of ideas but not to the ideas themselves. Thus, in the area of computer software, the courts have held that although the code of a computer program can be (narrowly) protected by copyright, its functionality can be protected, if at all, only by patents or laws governing trade secrets, not by copyright *(Sega v. Accolade 1993; Atari v. Nintendo 1992)*.

Further, in striking the balance between rights holders and the public, copyright law provides that reprinting portions of a copyrighted work for purposes such as criticism, comment, new reporting, teaching, scholarship, or research is a fair use and not an infringement of copyright. (We consider fair use as it relates specifically to trusted systems in a later section.) Additionally, provisions of the law (sections 108 to 120 of the Copyright Act of 1976) establish a framework—or, some might say, a patchwork—of more specific rights limitations, scope restrictions, and licensing arrangements. Many of these provisions are designed to address the concerns of particular interest groups, such as religious organizations, small businesses, lending libraries, blind and handicapped persons, cable television stations, and noncommercial broadcasters.

The history of copyright law has been, in part, a history of revisions intended to keep pace with changes in technology and media. In 1790 the copyright law governed only books and navigational charts; today it pertains not only to texts on paper but to all original works in a fixed, tangible medium of expression (e.g., motion pictures, architectural plans, sculptures, and sound recordings). Nevertheless, major revisions of the copyright law have been few and far between; almost seventy years elapsed between the 1909 and 1976 Acts. Even less-fundamental changes achieved through statutory amendment, regulation, or case law often take years and so lag far behind the latest technological developments.

This lag is especially apparent in the era of digital media, which change at an unprecedented rate. Moreover, digital media blur the boundaries between traditionally separate categories of work; thus statutory provisions intended to protect a particular type of work, or to serve a particular set of interest groups, can rapidly and unexpectedly become applicable elsewhere. For instance, is a web page that includes an animation and an audio stream an audiovisual work or a computer program? Or both? Or something entirely new?

Or consider a packet-switched computer network such as the Internet, in which the content-bearing information packets can travel either on cables or over the airwaves. The choice of a transmission path depends on moment-to-moment routing conditions that are unknown, even unknowable, to users. Should provisions of the copyright law pertaining to cable television or to broadcasting apply to the packets? Does it make sense to define a single concept of transmission for the digital media? And, if the answer is less than clear, should the law provide for a new exclusive right associated with all kinds of information transmission, as some have proposed? Yet another gray area involves whether the bits in an information packet traveling across the Internet are sufficiently "fixed in a tangible medium of expression" to constitute a work subject to copyright.

Digital media can also create confusion about who is the creator or owner of a work. Digital audio processing, for example, allows a composer to take digital samples from other (possibly copyrighted) works, process them in various ways, and include them in other, derived works. The source or sources of the sampled works may or may not be recognizable in their processed form; the samples can be very short, perhaps only a single drumbeat or a single note of music. In the absence of trusted systems, it is difficult, or impossible, to determine the original source of such a short sample. Even if the source can be identified as someone else's copyrighted work, it is not always clear (1) whether the person doing the sampling should pay the owner a royalty, and (2) what circumstances govern whether such borrowing of short segments constitutes fair use. Such digital sampling—or, in the absence of trusted systems, the lack of a precise method for controlling digital sampling—has already posed problems for the recording industry and recording artists. These issues have become even more complicated now that samples can be traded among thousands of musicians worldwide via the Internet.

Without trusted systems, effective enforcement of copyright in the digital media is nearly impossible. Like the effort to plug the proverbial sieve with its thousands of little holes, finding all the little infringement leaks of isolated individuals making copies is too hard and too expensive. Moreover, living with the leaks has its own deep risks. By publishing without copyright enforcement in a community that routinely makes unauthorized copies, rights holders accept the risk that, over time, such copying could become an established practice and could even be sanctioned by the courts as fair use.

In summary, the move toward digital media poses numerous challenges for copyright law and creates great uncertainty for rights holders, especially for would-be digital publishers. That uncertainty has hampered the adoption of the new media and discouraged most publishers from entering them. The impracticalities of enforcing copyright on untrusted, networked systems; the gray areas of legal interpretation for digital works; the lack of fine-grained control in copyright law; and the risk of an emerging legal claim of fair use for digital copying—all motivate aspiring authors and publishers in the digital media to find means other than copyright law for protecting their interests.

Digital Contracts

To consider one such alternative means, imagine a representative scenario of digital publication on a trusted system. The author begins by creating a work. When it is complete, the author finds a publisher to further develop the work and sell it (or perhaps decides to self-publish). The publishers develop a set of terms and conditions governing use of the work and, using a rights management language like DPRL, specify the time period over which the rights apply. They also determine what rights are applicable to the work: for example, whether printing is allowed, whether the work can be loaned out for free, whether members of a particular book club will receive a special discount, and so on. They may assign different fees for different rights: for example, deciding either to disallow creation of derivative works or to encourage creation of such works as a source of further revenue. Or they may require readers of the work to present proof—in the form of a digital certificate—that they are over eighteen. In DPRL, the statement of each right specifies the type of right, the time period in which the right is valid, the special licenses (if any) required to exercise the right, and the fee. In a trusted system, these rights are associated with the digital work either by bundling them together in an encrypted file or by assigning the work a unique digital identifier and registering the work and its rights in an on-line database.

Owning versus "Renting" Software

Having a specific and detailed agreement about the terms and conditions of a work's use is clearly an advantage to authors and publishers. But why would a consumer prefer such a system to the alternative: paying a single fee to purchase a copyrighted work outright and then using it in accordance with a set of legal standards applying to all digital works?

I would argue that specialized rules like those described above have potential economic advantages for consumers as well as for publishers. In the present software market, a customer typically purchases a software license for a fixed fee; someone who expects to make little use of the program pays the same fee as a person who will use it for many hours a day. In some markets, this situation is bad for both publishers and consumers, because low-usage consumers may decide not to purchase the software at all. In a trusted system with differential pricing and metered use, the amount customers pay for software would depend on how much they use it. Metered use would therefore allow consumers to "rent" the software under terms flexible enough to provide for decreasing unit costs for increased usage and conversion to purchase if their usage is great enough.

Another economic advantages might accrue to publishers and consumers of digital works through a variation on the first-sale doctrine. When consumers buy a paper book, they receive and own the copy of the book. After they have read it they are free to give it to a friend or to sell it. The first-sale doctrine in copyright law guarantees these rights. In the DPRL language, the analogous usage right is called a transfer right. Customers who purchase transfer rights can send the work from their own trusted system to a second trusted system; when they do so the copy on the first system is deleted or deactivated so that it can no longer be used. Like handing a book to a friend, a transfer operation maintains the same number of usable copies in circulation. The terms and conditions that allow a work to be transferred at no charge are thus analogous to those pertaining to paper books under the first-sale doctrine.

The right to transfer the work without a fee is exactly what a consumer intending to share it serially with others might want. On the other hand, from a publisher's perspective, a free transfer right is a threat to future sales. If all the people who read a digital work need to buy their own copy, the publisher will sell more copies. One solution would be to offer two different combinations of rights with a given work. In one combination, the consumer would pay the standard price and be able to transfer the work without fee—as in the first-sale doctrine. In another combination, the consumer could get a nontransferable copy of the work at a discount price, or later pay an additional fee to transfer it. Consumers who buy the work for their personal use and do not anticipate giving it away after using it might prefer the discounted purchase.

It can be argued that the first sale-doctrine is grounded in an experience with paper-based works that treats them as physical objects independent of their creative content. Like tools, household objects, or automobiles, such physical objects can be resold at the owner's convenience. Enforcing a law to prevent the resale or giving of paper books would be difficult in any case; so the first-sale doctrine makes practical sense. For digital works exchanged on trusted systems, however, these considerations are less relevant. The publisher and the consumer are free to enter into whatever agreement they see as economically advantageous.

Contract Law and Digital Contracts

In many ways, the terms and conditions specified in DPRL are similar to a contract or license agreement for using a digital work. However, although for convenience we refer to such a set of terms and conditions as a digital contract, it differs from an ordinary contract in crucial ways. Notably, in an ordinary contract, compliance is not automatic; it is the responsibility of the agreeing parties. There may be provisions for monitoring and checking on compliance, but the actual responsibility for acting in accordance with the terms falls on the parties. In addition, enforcement of the contract is ultimately the province of the courts. In contrast, with trusted systems, a substantial part of the enforcement of a digital contract is carried out by the trusted system. In the short term at least, the consumer does not have the option of disregarding a digital contract by, for example, making unauthorized copies of a work. A trusted system refuses to exercise a right that is not sanctioned by the digital contract. Over the longer term, consumers or consumer advocacy groups may negotiate with publishers to obtain different terms and conditions; but even then, the new digital contract will be subject to automatic enforcement by trusted systems.

Contract law is a complex subject that we cannot summarize here in any detail. Even so, we need to point out a few of the basic provisions of contract law as they relate to stakeholder interests and the design of trusted systems. First, however, we should note that at the time of this writing, there is an ongoing, controversial effort to add new provisions concerning "licenses of information and software contracts" to the Uniform Commercial Code, the body of statutory law that governs commercial contracts in most U.S. states. If the proposed provisions are adopted into law, they could have significant implications for digital publishing in general and for trusted systems in particular.

Contracts are agreements entered into by two or more parties. In a typical case, the parties negotiate and come to an agreement on the terms and conditions under which each party provides something of economic value to the other. This bargain for the mutual exchange of value is an important part of what makes an agreement a contract. Typically, a contract includes one or more promises backed by what are legally referred to as "valid considerations." The terms of the contract are usually set out in a written document, and the parties formalize their agreement with the terms by signing and dating the document. In cases warranting extra care, the parties' signatures may also be witnessed by a registered third party (a notary public), who asks the parties for proof of identity and may even take thumb prints or require other forms of personal identification.

A contract is backed by the force of law. Generally, if one party fails to comply with the agreed-upon terms and conditions, the other party or parties can enforce the contract through the courts. However, there are various circumstances under which the terms and conditions of a contract are not legally enforceable. They are preempted, for example, by the provisions of the U.S. Constitution and of certain statutory laws. The copyright law contains a preemption provision (section 301 of the 1976 Copyright Act) that may, in some cases, render a contract unenforceable. The courts have also developed various doctrines that hold certain contractual provisions to be unfair or improper (e.g., on the grounds of fraud, illegality, breaches of public policy, etc.) and thus unenforceable.

Here is one example of an unenforceable contract. It is not unusual for landlords to offer tenants standard rental agreements consisting of several pages of formal legalese. Suppose that a tenant signs such a document without noticing a clause in small print saying that if he eats mushrooms on Tuesdays he must pay an additional one thousand dollars in rent. The landlord throws a party on the next Tuesday and slyly offers the tenant mushrooms. Since such a clause is outside the normal scope of what belongs in a rental agreement, the courts would very likely refuse to uphold it. The courts thus provide checks and balances in contract law by deciding what contracts to enforce and how to interpret the terms and conditions of those contracts.

Checks and Balances in a Trusted System

With properly designed trusted systems, many of these same checks and balances can be available automatically. Consider, again, the case of the author who has finished a work and gives it to a digital publisher, who assigns it a set of terms and conditions. Like the conventional language used in legal contracts (so-called "boilerplate"), digital boilerplate in the form of templates and default conditions can be used to set up a digital contract.

Suppose, then, that the publisher has included some very unusual terms and conditions in the agreement. When the consumer's trusted system is connected to the publisher's trusted system, it first retrieves the terms and conditions of the digital contract and shows them to the consumer. Before the consumer accepts the digital work, his or her system can use a program (a "contract checker") to check for and highlight any unusual conditions in the contract (e.g., high fees for certain rights, unrealistic expiration dates, or other uncommon requirements). Because rights-management languages like DPRL are formal languages of limited complexity, simple grammar, and predetermined meanings, such a check is a straightforward matter for a computer. (As a somewhat bizarre example, consider a contract provision specifying that the consumer can copy a digital work for free but, surprisingly—and inconveniently—must pay ten dollars to delete it.) Before taking delivery of a work on a trusted system, consumers have the opportunity to agree to the terms and accept delivery of the work or to refuse them and not receive it. If a consumer agrees, his or her trusted system can digitally sign an acceptance form, which can be digitally notarized by a third party (a "digital notary") known to both parties.

The sequence of events in this example illustrates several checks and balances in the process. Both the publisher and the consumer can use computational aids to check the normalcy and appropriateness of the contract. More than a labor-saving or time-saving procedure, this approach is also a compensation for the intangible nature of information inside computers. It increases the confidence of both parties that the terms and conditions used by the trusted systems are reasonable.

One Digital Contract = Several Legal Contracts

It is helpful to think of a digital contract not as one contract but as a combination of several distinct legal contracts. There is the contract for access to the copyrighted work itself, and a second one for delivering the digital data, irrespective of whether the data are or can be copyrighted. If, for example, publishers use a trusted system to provide an uncopyrightable database *(ProCD v. Zeidenberg)* or a telephone white pages directory *(Feist v. Rural Telephone 1991),* they are entitled to charge for this service, even though the consumer could, in principle, get the uncopyrightable data elsewhere or put together his or her own database. Similarly, digital publishers, like paper publishers, could charge the consumer for delivering a copy of the complete works of Shakespeare, even though they are in the public domain. Like print publishers, digital publishers make life more convenient for the consumer, who must pay for this convenience. Publishers of an uncopyrightable work or one in the public domain cannot use copyright law to dissuade another publisher from offering consumers the same or a comparable work, as they could if the work were copyrighted. Thus, they must depend on trusted systems and digital contracts to maintain a business position.

Finally, there is a third contract implicit in the digital contract: namely, the agreement entitling the consumer to access the network of trusted systems in the first place. This agreement may be arranged between the consumer and the publisher, or between the consumer and one or more network service providers, who may or may not be affiliated with the publisher.

The idea that a digital contract includes multiple legal contracts provides a coherent rationale for enforcement of digital contracts, even those pertaining to uncopyrightable works. For example, suppose that a digital publisher provides a customer with a work in the public domain, such as the complete works of Shakespeare, under a digital contract that prohibits the copying or further transferal of the work in digital form. One consumer, however, is unhappy about this. He knows that the work is not protected by copyright and, when the bill arrives from the publisher, he refuses to pay it, or he sues to get his money back. In court, the consumer argues that charging for a work no longer protected by copyright violates the copyright principle of providing only limited-term monopolies for authors. Therefore, he claims, the digital contract should be preempted by the Copyright Act and held unenforceable. (The consumer might also argue that, because the publisher accepts the consumer's money while providing in return only a public domain work that ought to be available for free, the agreement fails for lack of consideration; i.e., because nothing of value has been delivered.)

The publisher responds that what is being sold is not the work itself but, rather, the service of delivering it. The publisher says, in effect, "Consumer, by dealing with me, you save time and energy and money over other delivery mechanisms, such as conventional bookstores. If I, as a vendor, want to provide this service to others, I must be entitled to collect revenue for the service, not just for the work itself. Therefore, it is legitimate for me to prevent you by digital contract from transferring the copy of the work I sold you." We think that the publisher has the better argument here. The consumer can still pay the publisher for the right to print out the contents of the book and can then copy the contents—for example, by hand or by scanning with an untrusted optical scanner. Moreover, other publishers can produce similar books containing identical texts, and a not-for-profit library can make these texts available for free. In short, the publisher has not overstepped the bounds of copyright.

Can the Consumer Negotiate?

Another point of possible concern with a digital contract is the extent to which a user can realistically negotiate the terms of the contract. In court cases challenging the validity of so-called shrink-wrap licenses—software contracts saying that by breaking the plastic seal on a package the purchaser agrees to the terms of an enclosed contract—plaintiffs have argued that such licenses give the publisher a power advantage and leave the user with only a "take it or leave it" choice. Many consumers do not bother to read the license. In principle, with trusted systems, it is possible to open a channel with the publisher to negotiate changes in such terms. It is worth noting, however, that one of the main advantages of digital publishing is the possibility of fully automated, twenty-four-hour shopping convenience. In that setting, renegotiating the terms of purchase for a mass market digital work is as unlikely as expecting to negotiate the price of buying a best-selling paperback at a convenience store in the middle of the night. The consumer would simply have to accept the terms offered or postpone the purchase until a human agent is available to renegotiate them.

Trusted Systems and Fair Use

In addition to specific statutory exceptions to the exclusive rights provided to rights holders, section 107 of the Copyright Act of 1976 sets forth four factors to be considered in determining whether a particular use of a copyrighted work is an infringement or a fair use.

1. The purpose and character of the use, including whether it is commercial in nature or for nonprofit educational purposes;
2. The nature of the copyrighted work;
3. The amount of the work used and its substantiality in relation to the copyrighted work as a whole; and
4. The effect of the use on the potential market for or value of the copyrighted work.

Fair use, itself, is not a public right. Technically, it is a legal defense that can be raised when a copyright owner challenges a particular use of a copyrighted work. In a representative case, the fair-use defense works as follows: A copyright owner publishes a copyrighted work. Without seeking permission, a second party obtains a copy of the work and incorporates portions of it in a new work. The copyright owner objects and takes the second party to court, claiming infringement of copyright. In court, the second party argues that the use made of the work constitutes a fair use and should be permitted. The second party might argue, for example, that the excerpt was used in a legitimate commentary or satire. The court has to consider the facts of the particular case in the light of all four factors. In practice, however, the fourth factor, the undermining effect on the market for the work, is often considered the most important.

One of the concerns some have raised about trusted systems is that they might block consumers' access to works they are entitled to use on a fair-use basis. Because a consumer could not extract a portion of a digital work on a trusted system, he or she would not have the opportunity to create the work that would occasion the fair-use defense. Of course, a trusted system would not prevent a user from manually retyping portions of a copyrighted book or from digitally recording excerpts from a copyrighted audio or video work through a computer's microphone or digital camera. It would however, preclude the operations of cutting and pasting excerpts directly from a rights-protected digital work unless the consumer of the digital work has contracted for the appropriate derivative-work rights. This example shows how far, in trusted systems, the pendulum has swung toward giving more power to authors and publishers.

Arguments about fair use for digital works sometimes tacitly (and incorrectly) assume that publishing risks in the digital medium are similar to those in the paper medium. However, while it is, as discussed earlier, unlikely that an infringer will make and distribute thousands of paper copies of a work, he or she can copy and mail a thousand digital copies with a single keystroke at no expense whatever. In other words, publishers who granted unrestricted access to each and every anonymous user on the basis of fair use would routinely risk the loss of all their copyrighted assets.

The other side of the coin is that fair use serves the public interest, particularly by helping protect freedom of speech. Parody, academic and social criticism, satire, and other forms of speech that rely on the ability to quote from, paraphrase, and modify portions of others' works are essential ingredients of the mix of speech that characterizes a democratic society. Fair use helps ensure that such borrowing can occur, even when copyright holders are vehemently opposed to the use of even short portions of their work in a critique or a lampoon. Without fair use, such rights holders could effectively quash criticism by preventing critics from publishing. The importance of this public interest aspect of fair use is amplified in the digital medium, where the ease of wide-area communications promotes the ability of everyone in society to engage in such free speech and be heard by all. Ultimately, this inclusiveness benefits society as a whole.

The Microtransactions Approach to Fair Use

Is there a way in the digital medium to balance the risks and benefits of fair use for publishers and consumers? One approach to the question argues that pricing and market forces will, in many instances, render the fair-use issue moot. Currently, fair use is a binary decision: either the use of a work is a fair use or it is not. When the courts rule that a particular use is fair, rights holders must forfeit the income from that use; if the use is declared unfair, the consumer must pay to use it. Thus there are high stakes in fair-use cases. The issue is especially awkward when the cost of being honest greatly exceeds the profit expected from using the work. In contrast, with trusted systems, fees for using copyrighted works in the digital medium may be either large or small and can be collected automatically. When market forces prevail, therefore, disputes that might otherwise arise over fair use will be resolved by setting fees at levels appropriate to specific uses.

Copying Music

A scenario about music illustrates how the market might resolve the issue of fair use. Suppose that a consumer wants to copy a short portion of a song from a friend's CD album. Today, consumers can easily make a tape recording for which the record company and the recording artist are not compensated. Such consumers are seldom found out by the record company; but even if they were they would probably argue that copying only a short part of a song is fair use. The company would very likely disagree. The point is that this is an all-or-nothing situation: either the consumer has to purchase the entire CD just to get a few seconds' worth of audio or the consumer pays nothing (because the copying is deemed a fair use) and the record company incurs substantial losses over thousands of such uses.

With trusted systems, though, a middle ground is possible. The fine-grained control offered by trusted systems makes it possible for consumers to purchase exactly the portion of the audio CD they want, and for the record company to charge and collect a fair price for that audio. In other words, to the extent that fair use is the copyright law's response to market failure, well-designed trusted systems can help correct that failure and eliminate the fair-use issue.

Copying Software

Another example of the microtransactions approach to fair use involves a person who buys a computer game or other software program for home use and expects to share it with other family members. In a household with several computers, he or she would probably make multiple copies of the software, one for each computer. Such copying today usually violates the shrink-wrap license that accompanies software. The family might assert, however, that their copying is fair use, because only one copy of the software is used at a time; in effect, they are simply sharing a single copy of the software.

In a trusted systems regime, would software publishers prevent this practice by requiring each family member to pay a separate fee for every use of the software? Such pricing practices would substantially increase the family's costs and provide it with no apparent benefit. Publishers would not necessarily take this approach. Even with trusted systems, they could keep prices at approximately today's level. They might, for example, sell software intended for household use with built-in digital contract provisions designed to make sharing the software among family members easy. They would allow a certain number of copies to be made free but would restrict their use to family members and their home computers. A family member who attempted to provide a "free" copy to someone not in the family, or to use it on a trusted system at work, would be prevented from doing so or required to pay an additional fee.

Printing Hard Copies

Yet another instance of the microtransactions approach to fair use suggests how the common practice of photocopying a single page of a book owned by someone else might change under trusted systems. Today, people typically make such copies without the publishers' knowledge. So long as the copying is fairly minimal, it is popularly considered (rightly or wrongly) to be fair use. With trusted systems, a publisher could discourage such unauthorized copying and, at the same time, benefit consumers. For example, the publisher could charge less for the right to view a page of a digital book on a display screen than for the right to print out a hard copy. This would benefit the reader who just wanted to look at one page rather than the whole book. The charge to view but not to print the page might even be made cheaper than the cost of making a photocopy.

A publisher might also charge less for the right to print the page in a form containing a machine-readable watermark than to print it without the watermark. The watermarked version of the printout would be less prone to unauthorized copying, because trusted digital photocopy systems would detect the watermark and charge for the privilege of copying it. Although consumers could still make copies on older photocopiers, the digital watermark would inhibit at least some unauthorized copying.

The Fair-Use License Approach

A second approach to fair use, one not grounded in faith in market forces or microtransactions, would institute fair-use licenses. In this scenario, Joe, a consumer, applies for a fair-use license the same way he might apply for a driver's or a radio operator's license. To earn the license, he has to study the rules of fair use and pass an examination by an appropriate organization, which we will call the DPT (for Digital Property Trust). The DPT then certifies Joe's identity and issues him a physical certificate as well as a personalized digital license to use on his trusted system. Under DPT, publishers of digital works would assign each work privileged rights that can be exercised by fair-use licensees. Publishers would also declare an insurance limit based on the expected commercial value of their rights to the digital work. Each time Joe exercises a copy transaction to obtain a copy of the digital work, the transaction fee includes an additional small amount—a share of the insurance premium on the digital work. Because of his fair-use license, Joe can exercise privileged rights to the digital work, but these privileged uses might be monitored, logged, and reported (subject to appropriate legal considerations for his privacy). Suppose, then, that Joe sends thousands of usable copies of the digital work to a mailing list on the Internet, and the publisher takes him to court, claiming damages beyond Joe's ability to pay. The court takes into account both the digital contract and the four factors of fair use. If the court finds in favor of the publisher, the fair-use insurance pays at least part of the damages. If the court finds in favor of Joe, the fair-use insurance pays for some or all of Joe's court costs and attorney fees.

The main point of this example is to illustrate that there are different risks and interests surrounding fair use in the digital media. In DPT, fair use is treated as a licensed privilege analogous to a driver's license, rather than as a legal defense. From a legal perspective, this is a substantial reframing of the fair-use concept that takes into account the greater risks of misappropriation in the digital arena. The example implicitly raises several interesting policy issues.

• Does Joe pay for his own license? If fair use is seen as essentially equivalent to free expression (a right), then fair-use licenses should be free, perhaps subsidized in some way by taxes or the publishing industry. If fair use is a privilege, like a driver's license, the potential licensee should pays a modest price for it.

• What rights does a fair-use license grant? In the example above, the publisher decides what additional rights go with a fair-use license. In an alternative scenario, there is a standard set of rights, possibly defaulting to zero-fee versions of all possible rights. Fair-use insurance covers the financial risks to the publisher if the work is turned loose on the Internet.

• Who pays for the insurance? In the example above, a per-transaction fee to pay the cost of insurance would be levied on all consumers of the digital work; this system would automatically scale the cost of the insurance to the popularity of the work. Alternatively, publishers could pay for the insurance, although this would amount to almost the same thing if a per-transaction fee passes the cost along to consumers. By using a per-transaction fee to collect insurance costs, self-publishers especially could avoid making up-front payment of the premium. In still another alternative, the fair-use licensee would pay the premiums.
• Whom does fair-use insurance protect? In the example above, it is intended to protect the publisher against losses. Should there also be a kind of liability insurance or fair-use bonding for consumers, to guard against claims for damages by publishers who accuse them of unfair use? Should there be a deductible on such insurance?
• Should fair-use actions be monitored? According to one view, such monitoring would violate privacy rights. Another argues that because monitoring would enable rights holders to detect violations of a fair-use license, it would be justified. An intermediate position would log and encrypt actions by the fair-use licensee and make these records available to law enforcement only when there is appropriate reason to believe that the law has been broken.

We have seen that the risks to a publisher in the digital medium for unencumbered fair use are much greater than they are in paper publishing. The first approach we considered to preserve the spirit of fair use in the digital medium relies on market forces and microtransactions to make most uses of a digital work very inexpensive. The second, which holds out the possibility of zero-cost fair use (and perhaps better safeguards the right of free speech), balances the risks and benefits by institutionalizing fair use as a licensed privilege backed by insurance.

A recurring theme in discussions of fair use and trusted systems is the fear that publishers will use trusted systems to take advantage of consumers by unfair pricing and that consumers will be unable to mobilize successfully against this practice. However, trusted systems must serve everyone's interests, or they won't serve anyone's. Publishers and consumers alike will be better served if publishers use trusted systems in a way that recognizes and responds to legitimate consumer expectations—for example, by creating digital contracts that preserve traditional notions of fair use. Publishers can either choose to self-regulate or risk being regulated by outside forces:

the legal system, the marketplace, and public opinion. Publishers who fail to consider consumers' interests may come under attack by free speech and civil rights organizations, consumer advocacy groups, and media commentators; they may find themselves the target of boycotts and a public outcry. Effective regulation may then emerge from legislative action. Moreover, digital publishers who get too greedy will find the competition offering better deals; or market forces may push back the entire industry. Consumers will simply continue to prefer works published in more traditional formats, and the market for digital publishing will remain limited—to everyone's detriment.

Accordingly, publishers who want to promote the growth of the market for digital publications will consider interests other than their own, and will make provisions for fair use. At the same time, as trusted systems make it easier for consumers to respect publishers' copyrights than to risk infringement and rely on the fair-use defense, our very notion of what practices constitute fair use will evolve.

Trusted Systems and Liability for Security Failures

Another issue of concern to potential digital publishers is who will be liable for losses incurred through security failures. Copyright law and the preceding scenarios for using digital works on trusted systems focus mainly on balancing the rights of two parties: rights holders and the public. However, because the trusted system technology has to ensure the security of digital works and enforce digital contracts, the manufacturers and vendors of such systems are central to preventing and dealing with the consequences of security failures.

A fiducial responsibility creates potential liability for trusted system vendors. What happens when the security arrangements for a trusted system fail and, without the action or intent of the user, a copyrighted document is released. Is the platform vendor liable? Or consider the case of an individual who purchases a digital work through a trusted system built by vendor A. That system is later used to transfer the work to a trusted system built by vendor B. Subsequently, the work is transferred to a third trusted system built by vendor C. This system fails in some way and releases the work onto the Net. Are vendors A and B liable because their trusted systems gave a copy of the work to vendor C's system, which proved untrustworthy?

The following scenario suggests one way in which the risks and liabilities for the failure of trusted systems might be handled. Vendor A builds and sells trusted systems that include hardware and software. Before bringing the system to market, vendor A takes the system to an independent testing organization, the DPT (or Digital Property Trust). The DPT tests the system, gives it a security rating, and issues signed digital certificates to be used by the trusted system in its authentication protocols. These challenge-response protocols and digital certificates make it possible for other trusted systems to determine the identity and security level of vendor A's system. They also make it possible to register all the transactions entered into by the system and, in particular, to keep track of which trusted systems have handled which documents.

Then, consumer Joan, using vendor A's trusted system, buys a copy of a digital work. Before offering the work on the Net, its publisher has assigned various rights to it, declared an insurance limit, and decided on the security level of trusted systems required to receive the work. Joan later transfers the work to a friend's system built by vendor B, and, as in the above scenario, the work is transferred again, to a system built by vendor C, which fails in some way (or, perhaps, is tampered with by an intruder); as a result the work is turned loose on the network. All the trusted system vendors, who are part of an industry coalition designed to deal with such problems, then take measures to isolate and limit the damage, and the publisher's document insurance evaluates and pays for the losses.

The example raises several difficult policy issues. Who is liable if a trusted system improperly releases a document because of a design failure? Who is liable if the security of a trusted system has been undermined by tampering or by a computer virus? Who is liable if a computer with outdated security measures participates in a transaction? Is it reasonable for publishers to take the entire risk, even with informed consent about the nature and limitations of the trusted systems? What prudent and appropriate actions might vendors take when a model of their system is apparently compromised? Should there be mandatory periodic testing and upgrading of trusted systems security? Should insurance rates be higher for works on trusted systems of low security than for those on high-security systems?

Over time, security requirements are likely to increase, and system failures are inevitable. When they occur, stakeholders in digital publishing will need to take prompt, concerted, and coordinated action to contain damage and maintain business as usual. The process of determining the cause of a failure can be complex. If the diagnosis and recovery take place in a highly adversarial atmosphere, trusted system vendors may have difficulty cooperating and sharing information in a way that facilitates expeditious containment of a security problem. Creating an industry insurance pool and establishing standards for cooperation can help publishers and platform vendors spread their risks and cooperate amicably.

In summary, the role a trusted system plays in enforcing usage makes vendors of trusted systems a party to the system for honoring intellectual property rights. The example above illustrates an approach that combines security technology with institutional arrangements. That approach is intended to create a business-as-usual marketplace in digital works in which risks are amortized by insurance, concerted and coordinated action by vendors is prompt, and the compliance and security of trusted systems is determined by an independent organization.

Governments as Stakeholders in Digital Publishing

As pundits have observed, international borders are nothing more than speed bumps on the information superhighway. The rapidity of digital communications and their ability to deliver information goods across national boundaries through trusted systems raises several issues of special interest to governments: import and export controls, national security, and taxation.

With regard to taxation, the automatic billing capabilities of trusted systems will almost certainly attract the interest of taxing authorities unless nations agree to treat the Internet as a free trade zone. In principle, of course, automatic billing of taxes is entirely feasible as long as trusted systems are kept abreast of changes in the tax laws of the trading partners.

More problematic is the issue of delineating boundaries in cyberspace. To a large extent, one computer looks like another in cyberspace, and location is not easy to determine reliably. Using intermediate agents, for example, it is possible to disguise the ultimate destination of a digital work.

Furthermore, with portable computers, the actual physical location can change continuously or frequently.

Consider this scenario. A Frenchman carries a laptop computer into the United States and downloads software from a U.S.-based software publisher. Has he exported it yet? Does he export it when he subsequently carries the laptop through customs? What if an American woman in France with a laptop computer she intends to take back to the United States logs onto the Net and downloads some U.S. software. Has she exported or imported software? Is she subject to French taxation? When she returns to the United States with the computer, has she exported the software twice or not at all? The traditional answers to these questions based on current law may or may not make much sense in cyberspace.

One approach to reframing import and export issues is to register computers in a way analogous to the way we now register ships or establish national embassies in foreign lands. If this approach is adopted, export would occur when someone transfers software from a U.S.-registered computer onto a French-registered computer, regardless of the physical locations of the two computers. Trusted systems could carry digital certificates that would authenticate their nation of registry no matter where they are in the world.

Although copyright law tends to be a national concern (though mediated internationally through the Bern convention), the digital medium intimately connects computers in different nations. The possibility of instituting automatic taxation on electronic commerce through a system of national registry is therefore likely to become a matter of great interest to various national governments.

Reflections

With trusted systems, copyright is potentially alive and well in the digital era, as is the balance of interests that copyright represents.

Trusted systems do not, however, exist in a vacuum. They are complemented by and are complementary to the legal, economic, social, and policy frameworks in which they operate. The Internet edge for digital publishing arises from the interactions among these frameworks and is driven by the desire of publishers and the public to utilize the Internet for commerce in digital works. The pushback comes from the need to balance the complex of stakeholder interests through copyright and contract law, market forces, and technology development. Another component of the pushback comes from traditionalists in paper-based printing who resist new business models and seek to impose print-based policies on the digital world. In this situation, chaos may arise from the legitimate, and often conflicting, interests of the stakeholders in digital publishing—authors, publishers, consumers, librarians, trusted systems vendors, financial clearinghouses, governments, and the public.

As we have seen, the pendulum is still in motion. Without trusted systems, digital publishing is at risk. If we have the wisdom to understand and institute the right policy choices, trusted systems may give us the leverage we need to guide the pendulum to an appropriate point of equilibrium.

5

Focusing the Light: Making Sense in the Information Explosion

The difficulty seems to be . . . that publication has been extended far beyond our present ability to make use of the record. The summation of human experience is being expanded at a prodigious rate, and the means we use for threading through the consequent maze to the momentarily important item is the same as was used in the days of square-rigged ships.

Vannevar Bush, "As We May Think"

Before the Internet, or even the widespread availability of digital computers, Vannevar Bush argued that society was creating information far faster than it could productively use it. Bush's 1945 observation—he was then director of the Federal Office of Scientific Research and Development, which directed the activities of over six thousand American scientists—led to the now-familiar metaphor, the "information explosion." Alvin Toffler popularized the phrase in his best-selling book, *Future Shock,* in 1970.

The word *explosion* sounds bad. People get hurt by explosions. Toffler used it to symbolize the difficulty he predicted we would have coping with rapid social and technological change and an overstimulating information environment. Information in the news, in our own areas of work, and even in entertainment is created far more quickly than we can consume it or make sense of it. For people drowning in information, Toffler believed, the explosion would be felt as a shock wave signaling arrival of the future.

According to historical accounts, of course, the information explosion is nothing new. We recognize that the volume of information available to people in the developed world has been increasing for hundreds of years. That perception comes from the ever-growing trail of written history, the growth of the literate population, and our easy access to information provided by technology—printing presses, telephones, radio and television, computer networks, and electronic information-storage devices. Individuals perceive the growth of information differently—depending on their personal and vocational situations and the responsibility they feel for keeping current.

A simple way to limit the explosion would be to stop creating information. But slowing down the output of information is impractical and seems wrong-headed. Would we really want to cut back the scientific publication that accelerates the search for the discovery and cure of disease? Or curb publication of the daily news? In Western democracies, limiting publication conflicts with a fundamental freedom, freedom of the press. Nor, sensing a popular cause, do we recommend curtailing the creation of movies, television, and other forms of entertainment.

The problem with the information explosion is not really that there is too much information. We already realize that we cannot know or read everything; we need, each of us, only to keep up with the documents relevant to our particular interests. In modern society people specialize and consume individual information diets. We each want a certain portion of information about the world at large, a certain amount about national and local matters, and a good deal of specific information about our circle of friends, our interests, and our occupations. The real problem with the information explosion is that it presents us with two dilemmas: being overwhelmed by useless information and having difficulty finding quickly the specific information we need.

Organizing the Information Soup

When Vannevar Bush considered the problem of the information explosion, he proposed addressing it with what he called a "memex" device: "It consists of a desk. . . . On the top are slanting translucent screens, on which material can be projected for convenient reading. There is a keyboard, and sets of buttons and levers. . . . Books of all sorts, pictures, current periodicals, newspapers, are thus obtained" (1945:107).

This early sketch of an information desk was extended by J.C.R. Licklider in 1961 and later described as a networked computer work station.

[The average person will have] his intellectual Ford or Cadillac—comparable to the investment he makes now in an automobile, or that he will rent one from a public utility that handles information processing as Consolidated Edison handles electric power. In business, government, and education the concept of "desk" may have changed from passive to active: a desk may be primarily a display-and-control station in a telecommunication-telecomputation system—and its most vital part may be the cable ("umbilical cord") that connects it, via a wall socket, into the procognitive utility net. (Licklider 1965:33).

Popular awareness of the information explosion has grown in tandem with the number of personal computers connected to the Net. The quantity of documents accessible by a personal computer—now reaching beyond the file system of one computer to systems all over the world—has increased by factors of millions.

One of Bush's influential ideas was to create direct links among the documents in the memex. Then, when someone reading an article comes upon a citation to another document, he or she could just press a button to jump directly to the article cited. Such linking is the defining characteristic of the hypertext systems developed in the 1980s and of the hypertext markup language that now permeates the World Wide Web. At its best, link-hopping is an efficient way to move between related articles and information sources.

Unanticipated by Bush was the explosion of publishing and self-publishing that now populates the Net. Because of this profusion, hopping across links through browsing or "surfing" is by now an impractical, unsystematic way to search for information. Although the average is skewed by the presence of index pages and big information sites, the average number of links leaving an individual web page is now about thirteen. Searching without a map can lead to interesting diversions but generally results in getting lost in cyberspace. Even completely automated web walkers, which hop across web pages at electronic speeds, now take days or even weeks to sweep through all the documents on the Internet.

Also implicit in Bush's approach was the assumption that the private organization of information by links would be augmented by enough librarians to help keep the world's knowledge organized. Here again, the proliferation of documents on the Net—which are constantly being changed, moved about casually, written, and deleted (and are usually unedited and unrefereed)—cannot be captured by existing library cataloging systems.

A useful technological method of finding relevant articles in the information soup of the Network is using the indexes and search services that retrieve documents according to the keywords they contain. Another response to the need to organize information on the web is the proliferation of pages listing links—sort of a "home brew" approach to particular topics. From a pragmatic perspective, these pages serve as fertile starting points for information foraging.

The Haystack Complexity Barrier

When we want to describe something difficult to find, we often use the metaphorical expression "as hard to find as a needle in a haystack." To get a sense of scale, I wanted to know just how difficult that task is. More specifically, I asked, how does finding a needle in a haystack—or, more generally, finding all of an unknown number of needles in a haystack—compare quantitatively with finding a set of relevant pages of information on the Internet?

There are many ways to search a haystack, and some of them provide analogies to our later discussion. One colleague suggested slyly that finding a needle is easy if you walk in the haystack with bare feet. Others suggested using magnets. If a haystack can be cut into parts, several searchers could use a divide-and-conquer strategy to search different substacks at the same time. In the simplest—and most arduous—approach, a person would pick up each piece of the haystack one at a time to see whether it is the needle or a blade of hay.

To answer the complexity question, I conducted a "field study" at the nearby Portola Feed Center. I assumed that hay in a haystack is packed at about the same density as it is in a bale. According to my simple observations, blades of hay in a bale are packed about ten to the linear inch and the average blade of broken hay in a bale is about eight inches long. Thus, a hundred blades of hay occupy a volume of about eight inches by a square inch and a cubic foot of hay contains about 21,600 blades. A cube-shaped haystack ten feet on a side would thus contain a little over twenty-one million blades of hay.

By comparison, near the end of 1997, the Internet contained about a hundred million web pages. If checking a blade of hay is comparable to checking a page on the Net, the Net was then about as complex as five haystacks. What surprised me about this field study was not that the Net's size had exceeded the "haystack complexity barrier" but that it had apparently done so in mid–1997. At the growth rate of about a factor of ten per year, searching the Net (albeit in an automated fashion) will soon dwarf this proverbially difficult search problem.

Information Feast or Famine

In the early 1990s, when browsers and search tools first appeared in the Internet, some of the researchers here told me I should consider finding another line of work because there would be no further need for librarians. But now they are calling for help in greater numbers, asking not only for documents but also for data, analysis, and ways to search the Web. People are overwhelmed by what comes back when they search the Net for information. It's feast or famine.

Giuliana Lavendel, 1997

Most people use a search service to search the Net for information. These services perform much of the time-consuming work before the searcher ever contacts it. Search services use web walkers ("search engines") to sweep through the Net periodically, following links and keeping track of the documents or web pages they have visited before. The web walker records the words used in each document visited, then assembles them into an inverted index, which records the documents on which each word appears. The index is saved and later used to quickly match customers' queries for documents. Thus, the time-consuming work of sweeping the Net and constructing the inverted index is done ahead of time and does not delay the retrieval of results when a user requests a search.

When someone searching for a document designates the words that should appear in it, the search service uses the inverted index to find matching documents. For example, the disjunctive query *haystack complexity* returns a list of web documents in which either the word *haystack* or the word *complexity* appears. The conjunctive query *haystack + complexity* returns a list of web pages on which both words appear.

The breadth of a query is an indirect measure of how many matching documents or web pages one would expect to find. A broad query matches more documents than a narrow query. For example, the disjunctive query *haystack complexity search* is broader than the query *haystack* because it matches more documents. Breadth may be quantified by counting the number of terms in a disjunctive query. For example, the disjunctive query *haystack search complexity* (i.e., documents containing the term *haystack* or *search* or *complexity*) has a breadth of three and would typically match more documents than the query *haystack search*, which has breadth of two.

Figure 5.1 shows how the number of documents retrieved from a large document collection varies with the breadth of a query. This curve reflects Giuliana Lavendel's "feast or famine" observation. For an example of the effect, suppose that we want to find a document that discusses the problem we are discussing here. Before writing this section, I connected to an online search service and started by asking for all documents that contain the phrase *haystack complexity barrier* verbatim. As this is a very narrow query, no matching documents were found. (This was not surprising, as I had invented the phrase while writing this chapter.) I then asked for all documents that contain that phrase or the term *search*. Over seventeen million matching documents were found, because the term *search* appears in so many places. To narrow down the results, I then limited the query, asking

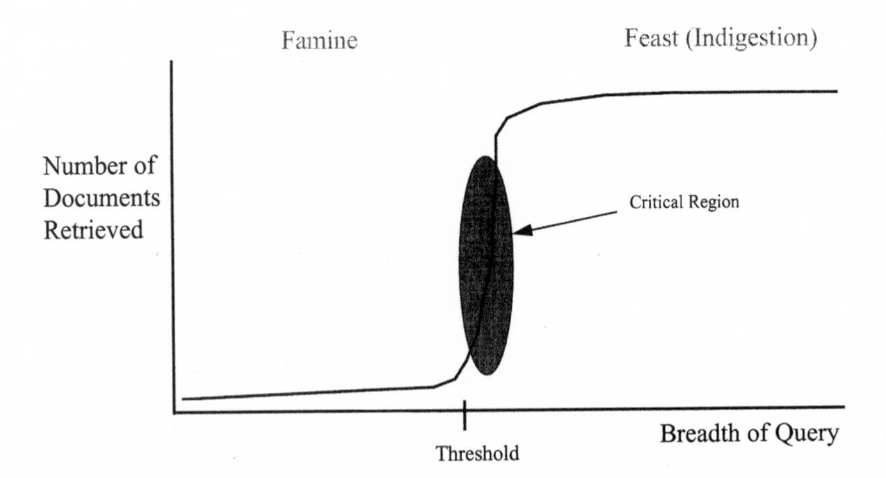

Figure 5.1
Phase Shift in Information Retrieval. For large document collections such as those found during an Internet search, a small change in the breadth of a query can result in a large change in the number of documents retrieved.

for all documents containing the three words *haystack, complexity,* and *search* in any order. The system found and ranked about fifty thousand documents containing at least one of these words. The top-ranked document was about managing complexity and included all the query terms, because it used the metaphor of finding a needle in a haystack. I then narrowed the query again by substituting the verbatim phrase *complexity barrier* for the word *complexity*. This change reduced the number of hits from fifty thousand to twenty-nine. The top-ranked document was an issue of an in-house magazine published by Digital Equipment Corporation about a line of computers. Thus, with only small changes in the query, the number of documents retrieved shifted from seventeen million to zero to fifty thousand to twenty-nine—fluctuating between feast and famine.

As suggested by this example, there are various ways to broaden or narrow a query by including certain terms or by requiring that they appear together, within a certain distance of each other, and so on. Short of running the query, we cannot determine the effects of particular changes with any precision.

This unpredictability and the extreme variation in the number of documents returned with small variations to the query indicate a phase shift in a search process; the two phases are the feast and famine in the number of documents returned. The term *phase shift* comes from physics and describes the sudden result of a small change, such as when a small decrease in temperature near the freezing point of water causes it to shift between liquid and solid phases. A point at which a transition occurs is called a *critical point*. A sharp transition from one phase to another at the critical point is called a *threshold effect*, and the region of rapid growth near a critical point is called the *critical region*. The shape of the graph in figure 5.1 is characteristic of a phase shift and is sometimes referred to as its *signature*.

Two standard measures of performance for information-retrieval systems are *precision* and *recall*. Precision is a measure of whether the documents returned by the process are relevant, and recall is a measure of whether all relevant documents are found. A deep problem of information retrieval is that near the critical region using a simple word-matching approach to retrieving documents forces extreme trade-offs between precision and recall. A broad query overwhelms the searcher with a flood of documents. However, though narrowing the query reduces the number of documents returned and increases precision, it sacrifices recall, and relevant documents may be missed. The problem is actually worse than that. Even at the right-hand side of the phase transition where the searcher is deluged with documents, many relevant documents may be missed owing to a mismatch between the vocabulary of the query and the vocabulary used in the documents. Thus, the conventional tools for retrieving relevant and useful information are deeply flawed.

The feast-or-famine signature of a phase shift in information retrieval is more of a concern for some sensemakers than for others. The casual browser may be satisfied with the results of a single probe if most small subsets of documents retrieved include at least one relevant document. On the other hand, professional sensemakers searching for rare information face the challenging and time-consuming task of crafting queries to steer their search. Sometimes it is not clear until we begin that we are looking for a needle, that is, that the information will be difficult to find. Recognizing that users feel overwhelmed by the feast part of the feast-or-famine problem, some network search-tool providers hide the problem by not showing them the large number of hits actually resulting from their search.

This precision/recall dilemma is a classic example of the difficulties of coping with complexity. Knowledge-based systems are computer systems that solve complex problems by using representations of knowledge to guide their search for solutions. In the design of knowledge-based systems, knowledge is the key to coping with complexity. What we need to effectively mine the critical region of figure 5.1 is a way to identify the relevant documents more exactly. Such knowledge for identifying documents would enable a search service to pluck out the needed documents without deluging the searcher with materials that match a query for accidental reasons.

Technology for Making Sense

In the late 1990s, the number of documents on-line grew by a factor of ten per year. This growth was fueled by the rapid start-up of new commercial and academic servers, the increased ease of posting web documents on popular service providers like America Online, and by the international expansion of the Internet. Even so, much on-line information—such as that provided by the on-line news services or posted on the growing body of private corporate intranets—never gets indexed. As digital rights technology is more widely integrated into the Net infrastructure, many more documents containing potentially high-quality information will be available on the Net for a fee.

Meanwhile, users' experiences with the feast-or-famine phase shifts of search services are creating a pushback from the Internet edge. Their frustration leaves many searchers feeling that the Net is an unreliable source. To the casual user of search services, the growth of information looks like an information explosion caused by network technology.

Although people may perceive technology as the cause of the information explosion, the most practical solution is also based on technology. It requires us to recast the problem from retrieving information to making sense of information.

Today's Leading-Edge Sensemakers

Within the increasing population of computer users there is a growing subset who spend a large fraction of their work life making sense of on-line information. These are people whose work requires them to sift through large quantities of data to understand something. Business analysts, who develop plans and strategic visions for new and established businesses, are sensemakers, as are analysts in government intelligence agencies and scientific leaders, especially those who work on multidisciplinary problems. Information specialists in libraries, policy analysts in think tanks and other information centers, and reporters and news analysts are all sensemakers. Trial lawyers looking for relevant cases are sensemakers. Patent attorneys looking for related patents and examining records of depositions are sensemakers. Students in college preparing reports on the material they are learning—as well as their professors conducting research in various fields of learning—are sensemakers. Even people who organize their e-mail into folders or devise bookmarks for the web are sensemakers. As they synthesize information from multiple sources, today's sensemakers have to extend their reach as more and more information becomes available on-line.

Sensemaking may seem like a solitary activity analogous to the lonely work of a scholar who spends years in the dimly lit archives of an old library. Yet sensemaking involves not only solitary individuals but also teams of collaborating problem solvers. The work of electronic sensemaking ranges widely—from small temporary points requiring rapid assessment to recurring and complex problems that may take weeks or months to solve.

Private and governmental research and consulting organizations often include departments or pools of experts with different specialties. When a client organization requests information on a topic, a task force drawing on specialists from the different departments is formed. The specialists compile information from different sources and try to make sense of it. Figure 5.2 shows a model of collaborative sensemaking. The process begins when a client requests a report on a topic. The first stage defines the task, establishing its scope and addressing any ambiguities in the initial request. A task force is then selected from the available experts. Next, the task force meets to identify potential sources of information and to brainstorm the questions and sketch out the organization of the report. During this part of the process, members begin to find relevant documents and divide the work.

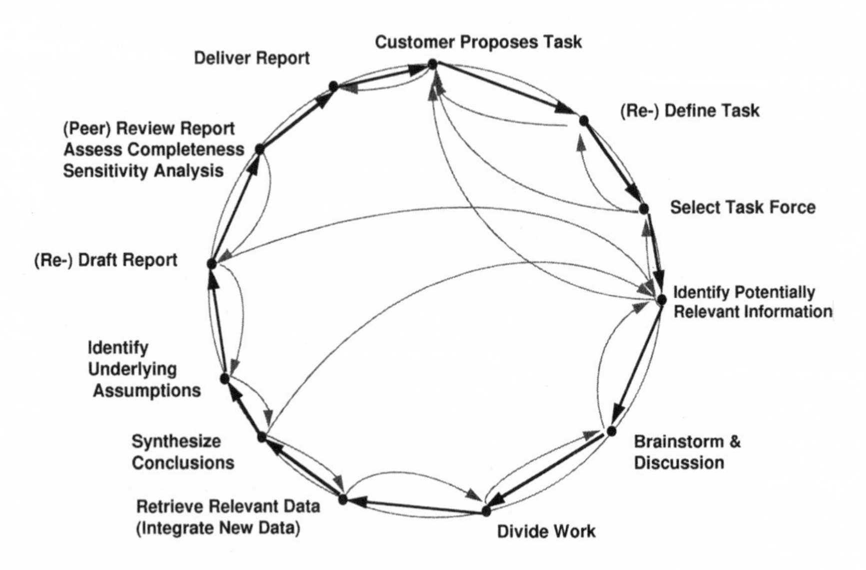

Figure 5.2
The Cyclic Work of Sensemaking. The graph illustrates the typical steps in a professional and collaborative sensemaking task, starting with a request by a customer who needs information and concluding with delivery of a report. The inner arrows in the cycle show that the task may involve feedback and looping. For simplicity, only a few of the many possible feedback loops are shown.

Once the work is divided, members of the task force begin to work in parallel on different parts of the task, pulling together bits of information and beginning to synthesize conclusions. Sensemaking about trends and generalizations often requires analysts to project the future and challenge each other to articulate sound assumptions on which to base projections. As the draft report takes shape, they may obtain peer reviews, which may challenge the report on the grounds of completeness or overdependence on unfounded assumptions. The cycle ends when the report is delivered to the client, although in many settings, such as the publishing of a newsletter, the cycle starts up again for the next edition.

The activity model of a collaborative team shown in figure 5.2 is not a perfect fit to all sensemaking situations. The work of any particular sensemaker or team of sensemakers depends on the social and institutional context. Some sensemakers work largely alone or in a loosely-connected network of colleagues. In a scientific setting sensemaking takes place when a group of colleagues and students turn their attention to writing a joint paper. A person writing a paper alone is also a sensemaker but collaborates with others only indirectly: by using others' published materials and in the peer review that takes place before publication. Some sensemaking tasks are so brief that they result in almost no written record. In other, more mature or seasoned sensemaking activities the explicit identification of underlying assumptions is a salient characteristic. The cycle described illuminates the range of sensemaking activities that are integrated into the reading, writing, and analysis of documents.

External Cognition

Psychologists who study how people work with knowledge or information pay close attention to their use of external representations: that is, to their writings and drawings on computer screens, paper tablets, or blackboards. They use the term *external cognition* for the human information processing that combines internal cognition with perception and manipulation of external symbols. We create, use, and share external representations in ways that augment our mental work.

I recently encountered a very effective example of a designed external representation when I was a member of a planning team preparing a proposal for a new business. We presented our proposal for review to a corporate oversight group using spreadsheets that set forth the financial aspects of the plan in a specified format. During the meeting, I was struck by the efficiency with which members of the oversight group flipped quickly to particular pages and stepped through the columns of numbers. Their probing of our plan followed well-practiced lines: Why do you believe you can hire people so rapidly during this quarter? Do these expense figures account for the staggering of employment start dates during the period? Why do you expect the income to climb so rapidly during this period? For them the formatted spreadsheet was a familiar external representation that made it easy to find certain kinds of information. Their effective use of the document depended on the way it grouped together exactly the information they needed.

In crafting external representations of a task a sensemaker typically gathers information from many places. Making sense begins with the selection and organization of the information to be used for making a decision. External representations help the group or individual both to figure out the problem and communicate the solution. In the process initial scribbles and informal notes evolve into a formal report. This point in the work invites a useful twist in terminology—referring to the external representation itself as the *sense*—the product of the analytical process. Viewed in this way, sense is not just an internal understanding. In writing a report or crafting a representation, sensemakers are literally *making the sense.*

Query-free Retrieval

The industries growing up around the World Wide Web have brought us many generations of browsers and integrated suites of tools. Yet the process of developing powerful sensemaking tools that scale up to the rapidly increasing amount of information available is still in its infancy. In this section, we consider several leverage points for producing new, more powerful generations of sensemaking tools—places where technology can make a real difference in augmenting human sensemaking.

On-line sensemakers often start with many terabytes of on-line information. However, even using external representations, sensemakers can see or manipulate only a few pages of information at once. The ability to leverage the power of computers depends on designing external representations that have powerful affordances for sensemaking (like the spreadsheets designed to visually present the strategic analysis of a business proposition). How can we design sensemaking systems that, by manipulating only a few pages of writing, can give us the computational leverage to make sense of terabytes of information?

One innovative approach to the problem is designed to retrieve information or documents without creating a query. In *query-free retrieval* a system for working on-line is integrated with an information-retrieval system. In one application developed at Ricoh Silicon Valley, a diagnostic system named Fixit is used to fix printers and copiers. Besides the knowledge-based reasoning stored in its software, Fixit has automatic access to a database of maintenance manuals. To access a manual, a technician using the system need not type in a query or refine it to retrieve information about a specific problem. Fixit "knows" the context the user is working in and can offer relevant references to those portions of the manuals containing information on the diagnosed fault in the particular type or model of equipment. The essential key to automatic query generation is thus a system that has a detailed map of the topic of interest and a way of discovering what a technician needs by monitoring what he or she is doing in the diagnostic process.

The help systems for programs on personal computers make use of the same basic idea. Query-free retrieval has also been used in several projects at Apple, where a computer system retrieves on-line information for users based on the work they are doing. Some information-retrieval systems use a particular form of query-free retrieval called *relevance feedback*, in which the system retrieves additional documents from a repository whose word-usage profile most closely matches a test set. In chapter 2 we described how a PDR system might generate automatic queries from highlighted phrases or digital ink markings that active readers make on a digital document.

Sense Maps and Snippets

Imagine that a sensemaker is assembling some notes on the telecommunications industry in preparation for writing a report. He or she might use any number of possible ways to organize the data—writing separate ideas and pieces of information in different regions of a page or structuring them with an outlining tool. Suppose that the sensemaker decides to make an outline to make sense of the various parts of the industry. The outline so far created is as follows:

I. U.S. Telecommunications Industry
    A. Industry Structure
        1. Regional Structure
        2. Forces for change
    B. Technological Trends.

Suppose further that while the sensemaker is editing the outline, the computer system is carrying out a search for relevant documents, using the outline to drive query-free retrieval. The system could open another window on the sensemaker's screen to list the retrieved documents, each of which relates in some way to the topics listed in the outline.

In the past few years text-processing systems capable of carrying out such tasks as writing document summaries have been created. They sometimes use the term *snippets* to refer to small chunks of text roughly the size of paragraphs. Snippets are bits of a document one might snip out with a pair of scissors. They are small segments on a single topic.

As the work of sensemakers is fundamentally compositional—sensemakers *make* sense—we have to ask: What are the units out of which they make it? Reports are too big for this purpose; sensemakers do not assemble an argument out of whole reports. The useful size for units of composition is the snippet. Consider, for example, the following portion of a snippet.

> The U.S. telecommunications industry is changing to create excess cellular and satellite capacity. This excess creates an opportunity for third parties . . .

We see looking back at the outline that this snippet seems most relevant to section I.A.2, which is about forces for change. How could the outline itself help us determine the most focused place for the snippet? Consider section I.A.2 again. Even without a more complete outline, we know from the structure of the outline that this section is not just about forces for change in general. It is about forces for change in the industry structure of the U.S. telecommunications industry. It is clearly not about regional structures or technological trends, which are covered in separate sections.

This picking apart of the outline into topics suggests a way to use its structure to target the mapping of snippets to the sense of the document. At any level of the outline the topic is essentially the conjunction (or logical additive-AND) of itself with the topics of its "parents" at all higher levels of the outline. As much as possible, entries at the same level should represent mutually exclusive (logical exclusive-OR) topics.

We use the term *sense map* to refer to an external representation that maps the snippets retrieved to the parts of an evolving sense document. As the sensemaker edits the sense—adding information, reordering the outline, or combining or splitting topics—the information search system is invoked, recomputes the set of relevant snippets, and presents them for possible incorporation.

In this way, the sense map is an artful way to combine the different compositional and retrieving aspects of making sense. In the community of researchers who have become interested in sensemaking, this combination is summarized by the following pseudoequation:

Sensemaking = Reading + Retrieving + Organizing + Authoring.

The equation says that the work of sensemaking is a process of finding and organizing relevant bits of information. Yet, although the electronic editors and browsers of the late 1990s are said to be integrated, they still split the work of sensemaking between separate tools for retrieving information and writing about it. And, working at the document level rather than the snippet level, they still require users to formulate queries.

Broadening Recall

The natural advantage of retrieving relatively long documents during a search conducted from a query accrues from the fact that most writers use a variety of equivalent phrasings to avoid producing a monotonous text. For example, an author may mention the *United States* in one passage, in another use the abbreviation *U.S.* or the adjective *American,* and in a third refer metaphorically to *Uncle Sam*. The retrieval of whole documents thus increases the chances that the words of the query will be matched to a relevant document.

This natural advantage does not apply so much to short document segments, in particular to snippets. Because snippet retrieval is less likely to find multiple phrasings of a concept, searches need to use techniques that match documents according to meaning rather than just words.

Semantic matching can improve recall for whole documents too. Some techniques reducing the requirements for exact word matching are already routinely applied in information retrieval. For example, most retrieval systems use *word-stemming* systems to remove suffixes and prefixes to convert terms to a standard form for matching; example, the words *dreamer, dreaming*, and *dreams* would all be converted to the root word *dream*.

There are several other ways of broadening the basis of a match. One is to look for synonyms. Thus, a search for documents including the term *city* could be broadened to gather documents using words like *town, metropolis, suburb*, and so on. There are, however, some difficulties involved in routinely broadening retrieval by using synonyms. Frequently, whether two words are synonyms or not depends on their context. As a kind of trick question, I sometimes ask people whether the words *man* and *woman* are synonyms. The usual answer—"Of course not!"—reflects our understanding that gender differences often matter. However, in discussions in which the issue is a common humanity or legal rights, the terms *man, woman, human*, and *person* are generally synonyms. For example, if we are searching for court cases about a man being robbed in a car, it is probably also useful to find cases in which a woman is robbed in similar circumstances. *Synonomy*—and meaning more generally—are context specific.

There are a number of relationships that can be used to broaden information retrieval. Suppose, for example, that someone is writing an article about mammals eating fish. It would be useful to include in the retrieval an article about a cat eating a fish, even though the term *cat* is not a synonym of *mammal*. As a cat is a kind of mammal, there is a class relationship between mammals and cats. Or, similarly, suppose that someone is writing an article about governments and the taxing of citizens and that somewhere out in cyberspace is a snippet commenting on court rulings on citizens' taxes. In this case, even though *court* is not a synonym of *government*, nor is a court a special kind of government, the snippet may be relevant. A court is, after all, an arm of government. Thus, there are a variety of relationships between terms that can be used to loosen the requirements for exact matching in information retrieval.

Increasing Precision

I was interested in knowing whether anybody ever found any viruses that attack the malaria parasite. I have all of Medline titles and abstracts since 1966. I can search them for strings—all the usual—and there's even a matching vocabulary or thesaurus that will do some level of translation for equivalence. The trouble is that the nouns are there but the verbs are elusive. I can easily find articles in which viruses are mentioned and malaria is also mentioned, but none of them have to do with what I'm talking about. I have no way to capture "viruses attacking plasmodia." There are so many synonyms for that and I just get hundreds of articles that are about coincidental infections.

Joshua Lederberg, 1997

I once had a conversation with the geneticist, Joshua Lederberg, about his use of information retrieval. He was looking at new approaches for curing malaria. In the late 1990s, malaria killed 2.7 million people each year, mostly children. A new generation of vaccines was being tried, but with only partial success. Malaria has evolved in a way that keeps it one step ahead of the body's immune response system, shifting forms and sites of infection from the bloodstream, to the liver cells, to red blood cells. Lederberg believed it might be possible to use a "counterattack" approach based on viruses or other infectious agents that attack plasmodia, the parasites that cause malaria.

To this end he tried to construct a query to locate such agents with an electronic search service. He knew that retrieving articles about viruses was far too broad. He also found that retrieving papers that mention both viruses and malaria was inefficient, because there were numerous articles about people with malaria who also had secondary viral infections. He was overwhelmed by the large number of irrelevant documents that matched his query for what he considered accidental reasons. What he wanted was a way to tell the search system to retrieve all texts that mentioned the two words *malaria* and *virus* in a particular relationship. That idea is at the core of an approach called *schematic search,* which is intended to make retrieval more precise. Lederberg describes the concept as follows.

I did a little—I wouldn't even call it an experiment—a very hasty trial run. But I reckon there are only about thirty verb contexts that I would need to formulate and it would essentially solve my problem. Think of all the major relational connections between nouns. You know, inclusion and exclusion, modification, subtraction—it

> doesn't take a great many of them. It didn't strike me as an impossible task to do this semantic conversion into a crude intermediate language. Perhaps that's a way that my problem might be solved. (1997)

Consider how this might work in the a less-technical example of a person studying the telecommunications industry. In this case, the sensemaker is looking for articles about non-U.S. companies buying a telecommunications company. Like Lederberg, the sensemaker wants to find documents or snippets in which a certain relationship holds among the words. Thus, the phrase "**Foreigners** *buy* telecommunications company" should match the following snippets:

MCI *to merge with* **British Telecom.**

NTT *considers buyout* of Motorola.

Siemens *increases holdings* in Deutsche Telekom.

In these examples, typographical differences indicate the parts of the snippets that correspond to the desired relationship among terms. Thus, the foreigners (boldface) in these snippets are British Telecom, NTT, and Siemens. The act of buying (italics) is expressed by the terms *merge, considers buyout,* and *increases holdings.* The telecommunications companies (underscore) are MCI, Motorola, and Deutsche Telekom.

The kind of semantic match in these examples is called a *schematic search*, where the initial phrase "foreigners buy telecommunications company" is used to create a schema that characterizes the required relationship. A schema indicates what kinds of terms or phrases can be used to fill in the relationship. Filling out a schema requires knowledge about the meanings of terms. Thus, to match the examples with non-U.S. companies requires knowing that British Telecom, NTT, and Siemens are the names of companies incorporated outside the United States. Matching the telecommunications company to the examples requires knowing that MCI, Motorola, and Deutsche Telekom are telecommunications companies. The relationship in this example is about buying a company. To match the examples requires knowing that merging, buying out, and increasing holdings are all ways of changing the ownership or control of a company.

It is possible to create computer systems that can work in the way these two examples suggest. Much of the research directed toward this development is taking place as part of the Message-Understanding Conference (MUC) sponsored by the Advanced Research Projects Agency (ARPA);

MUC evaluates empirical methods of extracting information from text. Such computer systems (or *knowledge-based systems*, as they are called) need to have encoded into them knowledge that enables them to perform the matches or, in the case of MUC, to extract information. This is precisely the kind of knowledge that would allow systems to work effectively in the critical region of the phase transition in the search process. Such knowledge is thus part of the critical leverage sensemaking systems need to cope with complexity and avoid the sharp trade-off between precision and recall—feast or famine—that plagues information retrieval.

The bad news, however, is that although encoding the knowledge for any given example is easy, the potentially enormous amount of knowledge of this sort needed to inform the search process for arbitrary topics makes this a massive task. The good news is that when semantic-matching knowledge isolates at least some of the pertinent articles on a given subject, the citations in the articles and in published indices can be used to locate related articles. Thus, given a starting point, related articles can be rounded up by following the citation links among the articles. The assumption that articles cite other articles works best for the scientific and scholarly literature. This is what happened to Joshua Lederberg when, months after he began his search, a colleague suggested searching for *Plasmodium* and *viruslike*. This netted a few articles, which then led to a treasure of other relevant articles.

Could the *viruslike* or *virus-like* term have been generated automatically by the retrieval-broadening process? There are many possible terms of similar meaning, including *viroid*, *quasi-virus*, *pseudo-virus*, *retrovirus*, *phage*, and *bacteriophage*. One search approach would treat generation and inclusion of these terms as domain-specific knowledge for broadening the retrieval; another would search the data base for terms that are variants of common roots. Both approaches have implications for system design. For example, proper use of such retrieval-broadening knowledge interacts with system elements for using synonyms, with the word-stemming system, and with other linguistic components of the matching software. Even if we solved the problem of how to generate a query using synonyms of *virus-like* as terms, we would find that including such terms in the search would intensify the precision problem by returning even more articles about secondary infections. In short, these approaches create a heightened need for semantic matching.

One approach to encoding the knowledge needed for schematic searching is to devise a system that allows individual sensemakers to encode such knowledge incrementally to increase the effectiveness of their work. The encoding for any particular sensemaker need only be complete enough to support the immediate task. However, because meaning is context-dependent, most sensemakers will need to maintain multiple minidictionaries of equivalences, perhaps even different dictionaries for different purposes. Furthermore, building a sensemaker's semantic-matching dictionary need not start from zero; he or she could begin with a small set of common relationships and then add to and tune it.

Attention Management

What information consumes is rather obvious: it consumes the attention of its recipients. Hence a wealth of information creates a poverty of attention, and a need to allocate that attention efficiently among the overabundance of information sources that might consume it.

Herbert Simon, as quoted by Hal Varian (1995)

Herbert Simon, the polymath economist and cognitive psychologist, has long studied how people use information in making decisions. Reflecting on the information explosion, he has often observed that human time and attention, not information, is the scarce resource. The biggest challenge is often how to allocate time to the most relevant information. From one perspective, this is just another corollary of his general notion of *bounded rationality*, the idea that people strategize to do make the best decisions that they can under the constraints of limited time and limited cognitive resources. As the quotation attests, there is a wealth of information but a poverty of attention.

Even in a sensemaking system that helps manage the information explosion by increasing both precision and recall, challenges to managing attention would remain. One impediment to incrementally building knowledge for semantic matching is that it requires sensemakers to perform two tasks simultaneously: making sense on the topic while tuning the search parameters—the semantic-matching knowledge of the system. The difficulty of doing so efficiently reflects the difficulties we have paying attention to several tasks at the same time.

Figure 5.3 illustrates the flow of snippets in a tool for sensemaking. As the user fills out an outline of the sense, snippets are retrieved from a repository and placed in the snippet arrival area. The shading of the snippets indicates how they map onto parts of the outlined sense. When a sensemaker selects a snippet, he or she can move it to the discard area (trash), to the snippet staging area, or to the sense.

An axiom of user interface design for collaborative tasks is that there should be a payoff to the user for any work he or she performs. When the user decides to discard a snippet into the trash, the system could ask for a specific a reason. Reasons for rejection might be, for example:

- *Bad source.* The sensemaker does not trust the information source. In this case, the system could ask the sensemaker to characterize the reason for rejecting the source in various ways, such as by the snippet's author, the news feed or publication that provided it, the person referenced in the snippet, or the snippet's genre.
- *Redundant data.* The sensemaker could have enough snippets of essentially the same kind, either because they are widely reported or because there are duplicates in the data base.

With the sensemaker's approval, the system could then intercept other snippets with the same characteristics and automatically toss them into the trash. Thus, as a payoff to the user for giving a reason to discard a snippet, the system provides greater assistance by discarding other snippets automatically.

In an ideal case, the snippet is directly useful and the user can incorporate it immediately into the sense document. In most cases, however, just copying the snippet into the sense document is a suboptimal approach. Usually it is better to encapsulate it into a digital object with other, similar meta-data. For example, the sensemaker could indicate that the snippet will be used as support for an argument, as a backup source or footnote to a section, or even to present a contrary opinion or counterexample. The value of such distinctions is that they enable the sensemaker to automatically analyze the completed sense report in terms of its use of data and sources.

The sensemaker may also decide to postpone action on a snippet. In this case, he or she could put the snippet into the snippet staging area with notations like the following:

- *Read later.* The sensemaker needs to read and think about the snippet but does not want to consider it right away; it has been selected as interesting but too complex for immediate use.

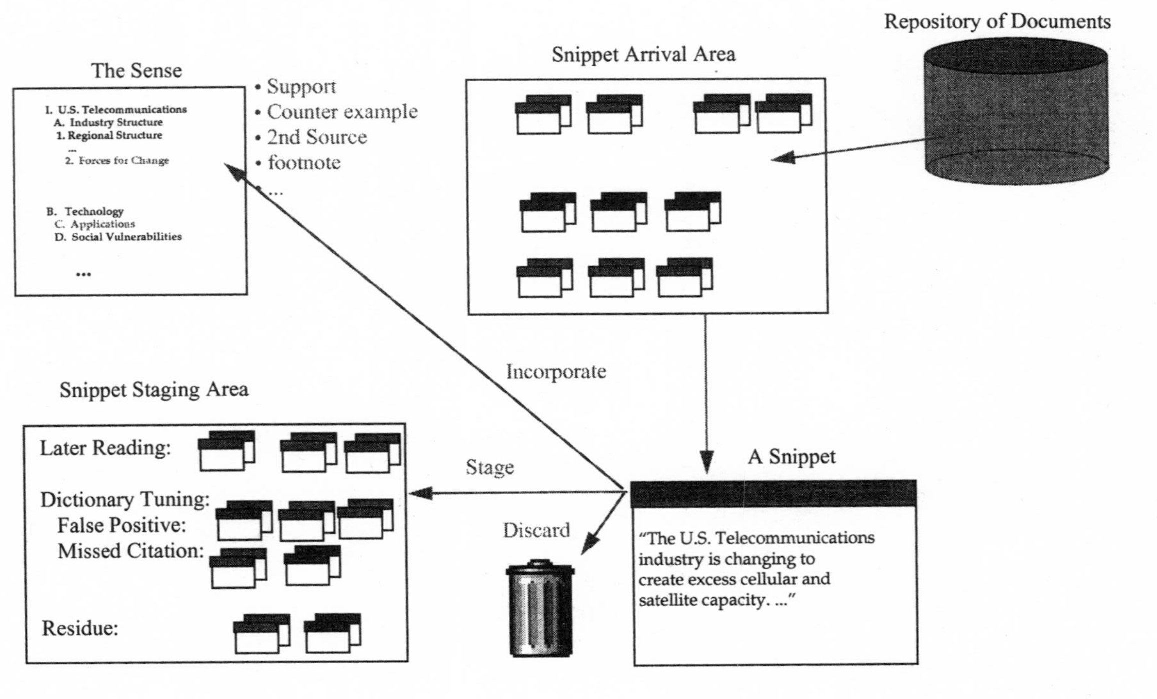

Figure 5.3
Managing Focus of Attention in a Sensemaking System. As the user fills out an outline of the sense (top left), the system searches a repository of documents for potentially matching snippets. The sensemaker can move a snippet to the discard area (trash), to the snippet staging area, or to the sense working area.

• *False positive.* This category means that the retrieval process has incorrectly collected the snippet. Saved as an example, the snippet can help guide or test the semantic matching by modifying the synonyms or other relations or fine-tuning the schematic search parameters.
• *Misfiled.* This category means that the snippet is interesting but probably belongs elsewhere in the sense document. Its appearance in the wrong location suggests that the sensemaker needs to tune the semantic-matching parameters of some other section of the report and use this snippet to guide the matching for it.
• *Residue.* The snippet challenges the basic categories of the developing sense and does not fit anywhere. The sensemaker needs to rethink the sense categories and then place this snippet where it belongs.

At the time of this writing, creating a sensemaking system like that described above is a-yet-unmet research challenge. Although the elements of such a system have been used in various information systems, they have not been tried all together in a system for sensemaking. Indeed, although the overall approach seems plausible, its effectiveness for sensemaking has not been demonstrated.

What can be said is that this proposal stands on fifty years of technology developed since Vannevar Bush first proposed the memex and addresses issues not then visible. It speaks in particular to how we might develop the knowledge needed to make the search for information more effective in the critical region. It also structures the overall task of sensemaking as an artful integration of reading, retrieving, organizing, and writing in a way that supports information retrieval from large document depositories without the need for formulated queries.

Reflections

The notion of bounded rationality causes one to reformulate what an information retrieval system should be in terms of benefit per unit time cost instead of precision and recall.
Stuart Card, 1997

The greatly increased amount of information now available on the Net—it has recently passed the haystack complexity barrier—has made the information explosion tangible for many people. Although thinkers like Vannevar Bush and J.C.R. Licklider anticipated the problem of the information explosion several decades ago, their solutions for dealing with it were never tested—because the large on-line databases needed to do so did not exist in their time. Meanwhile, our experience with such systems has revealed deeper issues in using large collections of information that they never anticipated.

Now, although people expect ready information from the Net, what they usually experience is information feast or famine. Often they cannot even determine whether the information they need is on the Net. They face a threshold effect, either finding nothing or being deluged with matching but useless documents.

Suppose for a moment that we possessed sensemaking systems like those described in this chapter. Would they effectively solve the problem of the information explosion and the threshold effect? In attempting to answer that question, we are in a position not unlike that of Bush and Licklider, because we don't yet have the sensemaking systems to try out. We can, however, learn from a thought experiment.

At the core of the sensemaking proposal is the idea that query-free retrievals can be generated from the sense document a sensemaker creates. Although this approach offers the possibility of great cognitive leverage—manipulating two pages to make sense of terabytes—it also contains the seeds of a possibly dangerous flaw. The system as proposed essentially works by first determining and then amplifying the sense the sensemaker begins with.

A familiar phenomenon occurs when a group of writers passes around a draft of an article they are writing together. We call it the first-draft effect, because the first draft of the document has such a great influence over the final form of the document. If the first draft is fundamentally wrong in some way or blind to some issue, then later drafts are likely to be defective in the same way.

The same danger exists in sensemaking. If the first draft or first sense is wrong or lacking in some essential, the system and further writing will tend to amplify the error. As the sense is flushed out, it can become more and more difficult to think outside of the box. Of course, this problem is not limited to machine-assisted sensemaking. Reflective analysts have seen such bias effects in individual and collaborative sensemaking in which there is no machine amplification.

Perhaps the root of the problem lies in the standard measures of information retrieval—recall and precision. Especially with regard to amplifying bias, using these metrics strictly contributes to the effect. Maybe what is needed is a greater appreciation of the value of outliers and contrary information. Imagine, for example, a retrieval system that returns snippets in three categories: relevant (mainstream), secondary, and outliers (contradictory). Indeed, we might even be able to develop automatic means of representing relationships among the snippets and using such relationships to generate suggestions for modifying the sense.

Another intriguing possibility is the idea that sensemaking systems could provide the basis for a kind of accountability of sensemaking. Once during a visit to an intelligence organization, I heard about a conversation that took place as a senior analyst was reviewing a draft intelligence report written by a junior analyst.

Senior analyst: How did you conclude that we would approve building oil pipelines through ______? [a middle eastern country]
Junior analyst: My source was a speech the Senator gave at ______. [eastern college]
Senior analyst: Don't you know that what senators say in such public addresses is for public relations and not policy?

The example suggests that analysts learn a lot about evaluating sources and using them for reliable sensemaking. In a similar way, tools for sensemaking systems could record the use and disposal of information from different sources and the reasons why it is used or not used. One plausible benefit of such tools would be that they would record not only the conclusions of sensemaking but also crucial parts of the process of *making* the sense. Such an "audit trail" could become the basis of a descriptive practice of sensemaking for teaching. The records could be used by junior analysts learning by example or by senior analysts mentoring junior analysts about the rules of good sensemaking. Moreover, like the outside auditors called in to check a corporation's books and certify that it has used good accounting principles, outside sensemakers could use the record to check a sensemaker's product to certify that he or she followed good sensemaking practice.

An interesting tension that arises from this example is that crossing the line from implicit to explicit rules of interpretation can be fraught with danger. Is the rule about public speeches valid for all public occasions? Does it apply only to senators, or is it about other public or private officials too? What exceptions are there to the rule? If the rule is not made explicit, then it cannot be acted on automatically, nor even passed on to colleagues easily. If a rule is only implicit, an audit of information potentially bearing on a decision would turn up sources ignored for no apparent reason. Clearly, formal sensemaking would challenge individuals and organizations to be explicit about their criteria and assumptions.

As we have seen before, the process of inventing the Net—including developing tools for finding and using the information on the Net—is also a process of shaping ourselves. We can design sensemaking systems that reinforce our biases, or we can devise ways to both leverage our access to information and challenge our interpretations of it. Sensemaking, like other uses of the Net, is a fundamentally social process. We have an opportunity to design not only technology but also appropriate ways of using it together.

The need for good accounting principles of sensemaking may become more crucial as more people rely more and more on the Net for information. A key, and potentially dangerous, characteristic of digital information is its intangibility and invisibility. As we increase the amount of information we obtain on-line, we risk becoming less familiar with and connected to the actual source of the information. Since we are not *there,* we are less able to use the clues and context of the situation to guide us in using the information. More than ever, we require expertise and care in combining information from multiple sources.

A good accounting practice for evaluating sensemaking could eventually become an important part of how society and individuals think about the effective use of the knowledge we generate and, especially, how to weigh our growing reliance on the Net to aggregate and distribute that knowledge.

6

The Next Knowledge Medium: Networks and Knowledge Ecologies

We are victims of one common superstition—the superstition that we understand the changes that are daily taking place in the world because we read about them and know what they are.

Mark Twain, *About All Kinds of Ships*

Tolstoy, one of the most interesting men who ever lived, explains that mystery of "interestingness" and how it passes from writer to reader. It is an infection. And it is immediate.

Brenda Ueland, *If You Want to Write*

Public opinion about artificial intelligence flips between extremes: "It will never work" versus "It might cost me my job." This dichotomy of attitudes reflects widespread confusion about artificial intelligence (AI), revealing a collective edge about the so-called smart technologies that are appearing more and more regularly in our lives.

In 1985 when I wrote the original version of this chapter as a thought piece for *AI Magazine*, AI was at the peak of a major growth phase. It had emerged from academic research laboratories, and images of AI systems and robots were gracing the covers of major news magazines. To develop AI systems, researchers were using specialized computers no faster than the outdated PCs that now lie discarded in the dusty corners of our schools and attics.

Over the next decade, the term *artificial intelligence* began to fade from public view. Conventional wisdom held that the "AI-hype bubble" had burst. The overblown expectations for smart systems were dashed, and an "AI winter" began. At the same time, the technologies of AI entered main-stream computing, moving onto the personal computers. Designers of conventional computers began adopting the object-oriented programming and representation systems that had powered the AI applications of the eighties.

By the late 1990s, sustained incremental progress in speech-to-text systems brought them to market in shrink-wrap boxes for personal computers. Manufacturers' assembly lines were automated by much-improved robotics technology, and individual expert systems were routinely used in many industrial applications. *Data-mining systems*, a spin-off of machine-learning research, were used to sift through on-line data to find informative patterns in everything from medical records to Shakespeare's plays. Haunted by the memory of earlier, unrealistic expectations and disappointments, computer scientists avoided the term *AI* even as the enormous economic potential of knowledge-based automation continued to power the AI dream. Meanwhile, other important elements of computer technology were moving forward to render AI applications more practical: computer speed continued to increase following Moore's Law, and the world accessible from cyberspace became more closely connected by the Internet.

Reconsidering the AI Dream

What is AI anyway? In the light of its short history, how can we think concretely about what it is, what it could be, or what it should be?

Most technologists are consumed with designing, building, and fixing things that need to work this year, if not next week. They can't spare much time for planning a more distant future. Nonetheless, those who are looking ahead believe that AI will fundamentally change our way of life. Although predicting the future is a notoriously unreliable process—at least as regards specifics—it is important to try to understand the trends and possibilities.

In this chapter we consider how AI technology, coupled with the Internet, could change civilization dramatically. The first part, "Stories," describes three examples of cultural change as it is studied by anthropologists and historians. The stories provide historical contexts for thinking about present and future cultural change. Next, to illuminate these stories and their lessons about technology, and to find appropriate analogies and metaphors for predicting the future, we consider several systems models from the sciences.

Finally, I offer several predictions, drawing on some current innovative projects and ideas that suggest how we might build a new knowledge medium—an information network with semiautomated services for the generation, distribution, and consumption of knowledge. Such a knowledge medium could quite directly change our lives.

Stories

The following three stories illustrate the constant growth of knowledge and cultural change in human history. They are representative; the anthropological and historical literatures contain many similar stories. Taken together, they demonstrate how cultures evolve, not steadily but in fits and starts.

The Spread of Hunting Culture

At the end of the Pleistocene glaciations, a spear-throwing hunting culture swept from what is now the northwestern United States and spread throughout the length and breadth of the Americas. This Paleo-Indian culture was characterized by the use of a spear with a distinctive fluted point and by the hunting of very large animals, such as bison and mammoths.

According to the archeological evidence, these weapons, and presumably the culture they were part of, advanced at a rate greater than a thousand miles a century. There is a debate as to whether the spread occurred through migration or through cultural diffusion when tribes living at the edge of a spreading cultural *wavefront* (*isochron*) observed, imitated, and integrated spear technology and hunting methods into their own life-styles. The weight of the sparse evidence, however, favors the migration of hunting bands seeking yet-unhunted herds of animals. In either case, the spear points illustrate how rapidly a prehistoric culture could spread over long distances.

The Spread of Farming Culture

The second story involves the diffusion of early farming culture. Using radiocarbon dating of the oldest farming artifacts discovered in the various regions, archaeologists have mapped the spread of farming culture from Eurasia and across Europe. Figure 6.1 shows a map of the land area of Europe, Asia, and Africa stretching from Germany and England southeast to the north shore of the Mediterranean and then south to Egypt. The wavefronts trace the progress of cultural diffusion from the Mideast across Turkey, Italy, Spain, Germany, and France at five-hundred-year intervals. According to the artifact record, farming reached Great Britain about four thousand years after it first appeared in Egypt. This diffusion rate is substantially less than a hundred miles a century.

It is easy to understand why the diffusion was so much slower for farming culture than for hunting culture. Farming is much more complex and requires practitioners to understand thoroughly several systems of knowl-

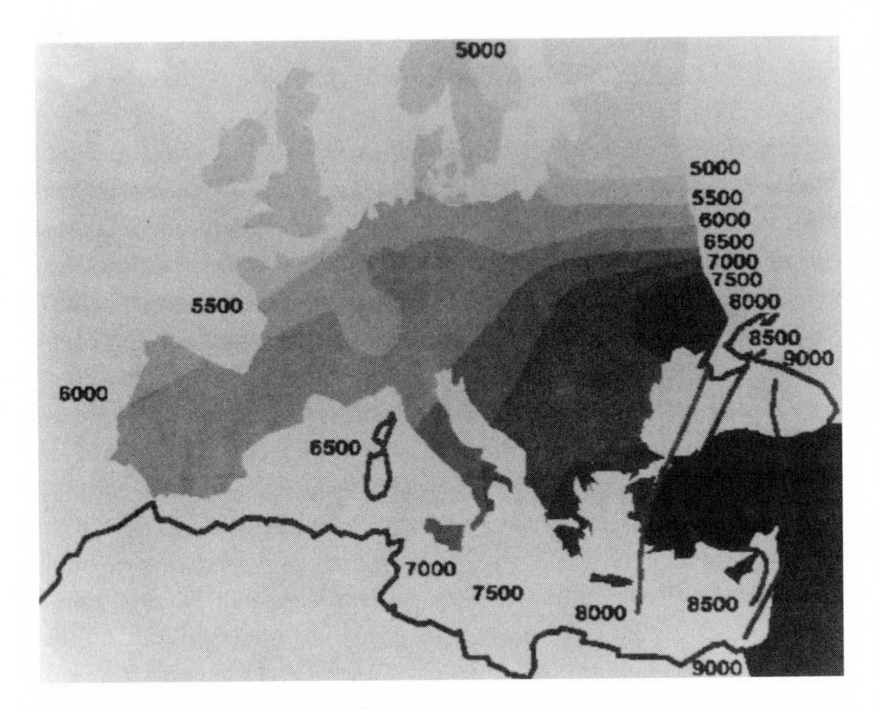

Figure 6.1
The Spread of Early Farming from the Near East to Europe. Wavefronts mark the locations of the oldest found farming artifacts in the archaeological record. (Reprinted with permission from A. J. Ammerman and L. L. Cavalli-Sforza, *The Neoloithic Transition and the Genetics of Populations in Europe.* Princeton University Press, 1984, p. 59.)

edge (about, e.g., seasonal changes, plant species, threshing techniques, animal breeding). This knowledge brought about some dramatic shifts in behavior and life styles—from nomadic food gathering to the more organized behaviors involved in planting, cultivating, and harvesting. Collective investments in social and work organization and (eventually) production of food surpluses led to great increases in population densities.

Peasants into Frenchmen

In contrast with the previous stories, the last example illustrates how a technology can accelerate cultural change in the modern era. It describes the expansion of roads and railroads in France between 1870 and 1914 and the sweeping changes that resulted from it. (See figure 6.2.) Similar stories of rapid change and modernization accompanying new communication or transportation technologies can be told for countries around the world, including the United States.

Until the mid-nineteenth century, France had only a skeletal highway system. Major roads led to and from Paris, the seat of the central government, principally to move armies, tax revenues, and food supplies to sustain the capital. The railway lines, which were begun in the 1840s, followed the same pattern. They did not connect the farms and villages with each other nor, for the most part, serve the everyday needs of ordinary people. Most of the real traffic continued to follow the trails, paths, tracks, and lanes that covered the countryside. Along these rough roads traveled the people, goods, and news of the time.

Until after mid-century many a peasants' world was restricted to narrow corridors—the space of a village and the familiar roads to local or regional market towns. Travel beyond the limits of a good hike was difficult and costly. According to custom, the few who went to Paris, even if only once in their lifetimes, were known thereafter as Parisians. During the winter the roads were so bad they were classified according to how deep a person or a horse trying to use them would sink in the mud—to the knees, to the shoulders, or to the head. In these seasons carts were unusable.

The exchange of many goods was limited to local villages and regions, which tended to be self-sufficient. Peasants selling wares at a distant market faced the prospect of carrying them for hours, then arriving at a town

Figure 6.2
Woodcut Print of the French Countryside, circa. 1890. Until the 1860s the highway system of France was a mere skeleton, and trade was limited mostly to neighboring villages. Travel was difficult and costly until an 1881 law mandated the building of rural roads to connect villages to each other and to the railroads. (Reproduced with permission from *Peasants into Frenchmen* by Eugene Weber. Stanford University Press, 1976.)

and having nowhere to store them. Once they arrived at the market, they were at a disadvantage, as buyers knew the peasants would have to get their unsold goods or livestock back home over the same unimproved roads.

Changes in the transportation system began between the 1830s and 1850s but rapidly accelerated after 1881 when, in the public interest, a law was passed to promote the building of rural roads. Together, the system of canals, expanded railroads, and interconnecting secondary roads brought new life to the villages, connecting isolated patches of countryside to larger, more distant markets. The roads connected the villages together, and the railroads connected the nation together.

As the geographical horizons of peasants expanded, they began to perceive new opportunities and to abandon traditional orientations. The peasants' apparent disinterest in trading on a large scale evaporated as the spread of public education gave them the necessary skills for shipping and receiving goods—reading, writing, and calculating—which took on greater significance as occasions in which to use them arose.

Productivity expanded enormously. A rule of thumb of the time was that economic activity in any area served by the railroad grew tenfold. In the Corrèze department of south central France, for example, improvements in transportation made fertilizer available. Between 1866 and 1906 its consumption increased by a factor of thirteen, and crop production increased sixty-five times over. Everywhere, industries were transformed as France began to function as a unified marketplace. Earlier in the nineteenth century the oldest of the cottage industries—spinning and weaving—had been displaced, first by water-powered textile mills, then by steam-powered factories; and coke had replaced locally produced charcoal in iron manufacture. Now, transported by trains to all regions of the country, factory-made cloth and metals displaced local products. The village blacksmith came into competition with nail-making machinery in distant foundries and couldn't begin to challenge the technologies that were fueling the enormous growth of the steel industry. Transportation expanded the marketplace and made possible what we now routinely call economies of scale.

The railroads and roads of France, like those of other continental nations, greatly accelerated the processes of change. In a scant forty years, they brought the French people a common market, a common language, a unified nation, and new prosperity. In the words of many French politicians, roads were the cement of national unity. As historian Eugene Weber argues, they transformed "peasants into Frenchmen." (See figure 6.3.)

Models

In the following subsections, we consider four models of systems and change from the sciences and social sciences: population genetics, ecology, economics, and knowledge processes in scientific communities. Although these models are well known, they are described here to provide

Figure 6.3
Stereoscopic View of Paris Transportation, circa 1895. By this time railroads and secondary roads were invigorating and nationalizing the French economy and supplanting river barges and horse-drawn carts. (Photograph by B. L. Singley, the Keystone View Company, Meadville, Pennsylvania.)

terminology, context, and metaphors we will find useful for making predictions about a knowledge medium.

Population Genetics

Beginning biology courses discuss the genetics of individuals, starting from the early plant experiments of Gregor Mendel to the more recent research that has revealed the chromosomal mechanisms of inheritance and the codon sequencing of individual chromosomes. The field of population genetics, on the other hand, goes beyond the genes of individuals to consider the variations and percentages of genes in populations. The set of genes in a given population is referred to as its *gene pool.*

In the model used by population geneticists, living organisms are gene carriers and a species is defined as a groups of organisms capable of interbreeding. Mutation occurs randomly in the genes of individuals; over time, reproduction and natural selection determine the percentages of different genes in a given population.

As the environment changes, the selection processes also change, and this is reflected in the gene pool. *Genetic drift* refers to a change in the relative distribution of genes within a population. A fundamental hypothesis of population genetics is that when two groups become isolated from each other, there is always genetic drift between the two gene pools; that is, the distribution of genes within the two populations begins to diverge. In general, the gene pools of larger populations have more stable distributions than those of smaller populations.

Sometimes new species appear and displace related species much more rapidly than would be predicted by apparent changes of environment or the expected rate of genetic drift. This phenomenon of speciation and displacement is called a *punctuated equilibrium*, because a population is stable for a long period of time but then experiences sudden changes. The fossil record might show, for example, that a change in the shape of a predator's teeth occurred suddenly.

The leading process model explaining this phenomenon posits three stages: isolation, drift, and displacement. First, a group of organisms becomes geographically isolated from the main population; second, it undergoes selection and genetic drift across multiple genes more rapidly than the larger body does. Finally, the geographic isolation ends and the slightly fitter group competes against and displaces the original population it was once part of.

Both the mathematics and the concepts of genetic drift and gene pools can be adapted to other systems, even nonbiological systems. Such systems need only have replicating elements and mechanisms analogous to genes for transmitting and recombining those elements.

Ecological Models

Ecology is the study of systems of organisms in an environment. A primary observation of ecology is that systems are composed of various interacting levels. This concept is perhaps best exemplified by food chains, in which big animals eat little animals and so on down to the most rudimentary plants and microorganisms.

Levels are a simple model of relations; in complex ecological systems, the relations among species form an intricate web. To describe these relations ecologists have developed a rich vocabulary of terms: for example, *predators, symbiotes, parasites,* and *pollinators.* Each species in a system has its own *ecological niche;* collectively they fill all the nooks and crannies needed to make the ecology function efficiently.

When several species evolve together in ways that increase their mutual adaptation, they are said to *co-evolve.* From this mutual adaptation can come increased efficiency, which leads to an important ecological principle: *life enables more life.* Everything depends on everything. This degree of interdependency does not mean that ecologies are fragile. They are not constant, nor are they formed all at once. They develop under processes of co-evolution as multiple species compete for and create niches. As an ecosystem increases in complexity, it also becomes more redundant and thereby more robust.

We can characterize other systems with populations of replicating elements in terms of metaphors drawn from population genetics—for example, drift, mutation, and selection—and from ecology—for example, ecological niches. All these metaphors describe relations between particular groups of elements and the way they co-evolve.

Economic Models

Economic systems are similar in many ways to ecological systems. Businesses form economic subsystems and, like species in an ecology, depend on each other. There are suppliers, distributors, and consumers. Subcontractors supply parts and services to multiple manufacturers. Corporations and goods are said to occupy economic niches. New products can drastically change the shape of a market by displacing older products from existing niches or by creating new niches.

Economics brings us several concepts not found in ecology, including *price, supply,* and *demand*. These concepts provide a quantitative basis for explaining action in the marketplace. A market is said to "seek its own level" according to the laws of supply and demand. When there are many suppliers in a market, some are more efficient than others and consumers benefit from lower prices. Such effects can ripple through an economy when, for example, the price of a part used in many different products is lowered. When there are many consumers, producers can often achieve economies of scale by switching to large-scale manufacturing processes and mass production.

Thus, business enables more business. This can also be seen in the proliferation of businesses. The first businesses in a rural area are basic and relatively inefficient, but as a locality develops a rich mix of more efficient businesses, its economy becomes more stable and robust.

Some economic systems require more sophisticated models than the essentially laissez-faire ideas just described. For example, when an economic system interacts with a legal or regulatory system, the model may need to explain more subtle phenomena. Nonetheless, the concepts described here delineate a reasonable first-order model for many situations and are enough for our purposes.

Social Processes in the Scientific Community

Anyone with even a casual familiarity with science has heard of the scientific method. It is a formal means to create and validate new knowledge. Studies of the actual conduct of science, however, reveal a social dimension to the conduct of science that goes beyond the scientific method.

Scientists are knowledge workers who have important professional relationships with each other. Such relationships are intrinsic to the practice of peer review and the "invisible colleges" of colleagues who collaborate and often share their results prior to publication.

Scientists also fill different niches in the ecology of science. Some are known as innovators, and others excel at integrating the results of others, or at reliably pursuing crucial details. Some are best at theory, while others are gifted at overcoming difficult challenges in experimental design. Yet others are good at explaining things and make their major contributions through teaching.

Thus, scientists have many different roles as knowledge workers. Science enables more science, or, perhaps knowledge enables more knowledge. Yesterday's discoveries and unanswered questions drive today's experiments and provide the backdrop against which they are carried out. This additive effect is particularly evident in practical knowledge about the techniques of experimentation. For example, a series of experiments about genes and nutrients might yield a well-characterized culture of microorganisms. This culture might then be used for fine-grained studies of genetic exchange. Their results may, in turn, enable other scientists to conduct experiments about the mechanisms that regulate gene expression.

Predictions: Toward a Knowledge Medium

> We say that some piece of information is beautiful when we experience a particularly deep emotion as we absorb it—when we look at it, read it, hear it, or feel it. We feel this emotion when knowledge has been transferred with power. I have little doubt that this feeling, which is so wonderful that we wish to experience it over and over again, is genetically based, for it is tightly intertwined with our existence as a highly communicative species.
>
> Rich Gold, "No Information without Representation"

In the last chapter of his book *The Selfish Gene* (1990), Richard Dawkins suggests provocatively that ideas (he called them *memes*) are like genes and that societies have meme pools just as they have gene pools. The aim of Dawkins's work in biology was to shift attention toward a "gene's eye view" of the processes of evolution. Although when mammals reproduce they do not (yet) clone themselves, and offspring are not identical to parents, the latter do pass along their genes—the mostly invariant units of inheritance. Looking at issues from this point of view goes a long way toward explaining many of the persisting conundrums about traits and behaviors that are genetically linked.

Although they were almost an afterthought in his book, Dawkins's memes have been taken up by many writers. Memes, somewhat like genes, are carried by people. They are the knowledge units transmitted in conversations and contained in minds. Memes can be reinterpreted in new environments and expressed in new combinations, just like genes; and, like genes, they often come in interactive clusters. Memes compete for their place in meme pools and, if populations of memes become isolated, they undergo a sort of memetic drift analogous to genetic drift.

As Rich Gold points out in the quotation above, we often experience a deep emotion when we absorb a piece of information—a meme. We may understand a meme about how to do something when we see someone else do it, or when we read or hear something. The "ah ha!" feeling inventors get at the moment of insight is a variant of the same feeling.

In the light of this feeling, we can reinterpret the slogan "knowledge is power." When we learn something important, that uplifting feeling may reflect a little surge of power. If Gold's comment is correct, it suggests something deep about the survival strategy of the human species: that we carry genes—curiosity genes, teaching genes, and learning genes—that promote the replication and selection of memes. Thus, we are wired to enjoy those activities that have great survival value.

Toward a Meme's Eye View

We can reinterpret the preceding stories of cultural change from a meme's eye view. Cultural change occurs along a wavefront, with the memes competing and spreading to new carriers. Basic human capabilities for communication and imitation modulate the rate at which the memes spread. Differences in the rate of propagation, like those between hunting cultures and farm cultures, can be explained by assuming that many more memes are needed to communicate farming practices than hunting skills. Apprenticeship programs and university courses in science can be seen as social mechanisms for communicating memes about the techniques and practices of science.

The progression in our stories from hunting culture to farming culture to the modernization of France is one of increasing cultural complexity. It is not, however, a sequence of decreasing speeds of propagation. Considering only the complexity of the cultural shift, we might expect that the nineteenth-century modernization phase in France would have taken many centuries, if not millennia; nonetheless, in spite of the dramatic cultural changes that had to take place, the rate of meme propagation sped up enormously in the late nineteenth century. To understand this, we must reconsider the effects of French road building.

The roads did more than change France into a national marketplace for goods; they also transformed it into a marketplace for memes. The isochron waves, which so faithfully described the orderly flow of memes for the hunting and farming cultures, are completely inadequate for tracing the flow of ideas along the roads and railroads of France. Technology changed the process. Imagine the memes crisscrossing France along the roads and railroads, creating an intricate pattern of superimposed cultural wavefronts. Ideas from faraway places were continuously delivered, reinterpreted, and reapplied.

By bringing previously separate memes into competition, the roads triggered a shift in equilibrium. The relaxation of constraints on travel led some to meme "displacement." Cottage industries were replaced by mass production, and the way of life changed. Multiple equilibria were punctuated at once. The very richness of this process accelerated the generation of new recombinant memes with their own wavefronts. Whole systems of memes (e.g., how to run a railroad station, what the value of education is, even how to speak standard French) were created and transmitted. As Weber noted, peasants became Frenchmen in a mere forty years.

Since 1914 several new communications media have been introduced, including, besides improvements in the post office and telephones, television and the Internet. These communications media have quantifiable properties that govern the transmission of memes: transaction times, fan-out, community sizes, bandwidth, and storage. Better post offices mean that people can spend less time traveling and have more time for other activities. They can order goods by telephone or over the Internet and receive them by overnight mail. The rise of the mail order catalog stores at the turn of the century was a manifestation of this change. The strike at United Parcel

Service in 1997 showed how many sectors of the economy depend on overnight delivery of packages: just-in-time delivery for demand-driven processes. The catalog stores of the United States are now as important for business as they used to be for rural communities. They make it possible for suppliers to reach large areas and for telecommuting workers and rural-based software companies to take delivery of specialized hard goods conveniently, even if they live far from urban areas.

AI Technology and the Internet: Not Yet a Knowledge Medium

Precisely defining a knowledge medium is much like defining life; and, like life, it is better characterized in terms of processes than of properties. Life is usually described in terms of such processes as reproduction, adaptation, growth, and food consumption. Similarly, a knowledge medium is characterized in terms of processes such as the generation, distribution, and application of knowledge and, secondarily, in terms of specialized services such as consultation and knowledge integration.

There are many borderline cases that defy a simple definition of life. Fires spread, change their burning patterns, increase in size, and consume fuel, but we do not consider them alive. On the other hand, viruses and plasmids are classified as living because they take over the life processes of their hosts, even though they themselves lack the machinery for reproduction. It can be said that mammals are more alive than viruses because the quality of their processes is so much richer.

Knowledge media also have such borderline cases: for example, communications media without knowledge services and databases with limited distribution and services. Just as life is thought to have come from things that were nearly alive, so too might genuine knowledge media emerge from borderline media.

AI research includes many topics relevant to knowledge media: the representation of knowledge in symbolic and subsymbolic structures, the creation of knowledge bases for storing and retrieving knowledge, the development of problem-solving methods that can use and be advised by knowledge, and the creation of knowledge systems (or expert systems) that apply knowledge to solve specific problems (e.g., IBM's Deep Blue or computer-based diagnostic systems). However, even with the Internet, AI technology does not function in an important way as a knowledge medium in our society. Its influence has been far less important to the creation and propagation of knowledge than the secondary roads were to the modernization in France.

This is more than a matter of the youth of the field. The main goal of AI has led off in a different, possibly contrary, direction. The term *artificial intelligence* expresses what we most commonly understand as the goal of the field: creating intelligent, autonomous thinking machines. Building such machines brings to mind ideas quite different from those we associate with building a knowledge medium. It suggests the creation of machines that are intelligent and independent of our control. In contrast, the goal of building a knowledge medium draws attention to the main source and store of knowledge in the world today: people.

As it turns out today, nobody really knows the form that a knowledge medium will take, or how to use the Net and AI technology as a knowledge medium. There are roughly three main ideas or approaches that I know about that may be helpful. For simplicity, I refer to them as parts, services, and documents. The essence of these ideas is rather simple, but the practicalities and issues are more subtle. These ideas are explored in the rest of this chapter.

Books versus Expert Systems

In the meme model, a carrier is an agent that can remember a meme and communicate it to another agent. People are meme carriers, and so are books. However, there is an important difference: people can apply knowledge, whereas books only store it. Librarians, authors, publishers, and readers are the active elements in the printed knowledge medium. Computers can apply knowledge as well, and this ability makes them important for creating an active knowledge medium. When a medium includes computer systems, some of its knowledge services can be automated.

The most promising automated knowledge processors today are *expert systems*. In several well-publicized cases, expert systems have proven to be of substantial economic value, far greater than the cost of their development. Expert systems are computer systems that solve narrowly focused problems at an expert level. They are built by encoding the rich knowledge of experts into computer memory—either by a painstaking process involving interviewing experts, a process of statistical learning, or a combination of both.

Although tools for building expert systems and for programming in general have continued to improve, designing an expert system is quite different from writing a book. In writing a book, an author needs to get the ideas together and to write them down clearly. Sometimes connecting ideas will be missing or out of order or slightly wrong. But authors depend on the intelligence and knowledge of their readers to understand and integrate what they read. Not so with computers and expert systems. Today's computers are less sophisticated than humans and, probably, many species of animals.

Knowledge must be acquired, represented, and integrated when programming an expert system. For communication to be effective, the programmer must develop and provide the computer with a common ground of language and concepts within which to integrate the knowledge presented. In any community of experts, there are variations in knowledge and practices that must be discovered and examined. Beginning with observations and interviews, the process for building an expert system moves the experts' knowledge through a series of informal, semiformal, and formal representations. It then cycles through steps for documenting, codifying, implementing, and testing. In the current state of the art, the underlying tools used, although they assist in the construction of expert systems (just as text editors assist authors), provide no memes of their own to help organize new knowledge or to fill in gaps. All expert systems require their knowledge base to be handcrafted, which is why they are so expensive. (Readers interested learning about the practices of acquiring, representing, and testing knowledge in expert systems can find a detailed discussion of the art in chapter 3 of my textbook, *Introduction to Knowledge Systems* [1995].)

Specialized Production of "Parts"

Mass-producing complex artifacts like automobiles does not rely on such extensive handcrafting. What is sometimes referred to as an *operational economy* enables carmakers to exploit a marketplace for ready-made materials and subassemblies. They do not need to make tires and batteries, or mine metals, or produce glass and plastics. They are not interested in mastering the details of all the necessary technologies but do want to exploit economies of scale in the market. Specialized companies can produce batteries, glass, and tires less expensively than automobile manufacturers can, thus making it possible for them to build complex goods that would be infeasible if everything had to be made from scratch. This fact brings to mind the abundance rules of our models: life enables more life, business enables more business, and knowledge enables more knowledge.

The analogous goods of a knowledge market are the elements of knowledge or, if you will, the memes. But in today's expert systems it is still necessary to build them from scratch, meme by meme, into the knowledge base of each expert system. These are the "parts" of specialized knowledge production.

Building expert systems, therefore, is more like writing books than building automobiles. Both are highly creative enterprises, both require research to collect the facts, and in neither case is there any (or any appreciable) economy of scale. However, compared with the number of people who are skilled in using the printed medium, knowledge engineers are few in number. They are the computer-literate monks of the twentieth century illuminating their manuscripts in splendid isolation, awaiting, perhaps, the invention of the next printing press.

Standard Terms and Meanings

As the foregoing discussion makes clear, one way to reduce the cost of expert systems is to build them with knowledge acquired from a marketplace. Doing so requires us to set in place some new processes and make some crucial technical advances. How might we do that?

The technical issues are not just the usual problems of electronic connection. Millions of computers are now connected on the Internet. They are routinely used for e-mail and net surfing and, increasingly, for electronic commerce. The networks carry mostly data, not knowledge—low-level facts, not high-level memes. The precise distinction between data (or information) and knowledge is elusive, but its general sense is that very little of what computers are transmitting is akin to what people talk about in serious conversation. Thus very little of what is transmitted can be used in a direct and substantial way by the computers.

Imagine how we might draw on a collection of knowledge bases from expert systems. These knowledge bases, although developed for different purposes, would have some important terms in common. For example, consider the term *water*. A chemistry-knowledge base would specify when water freezes and boils and what dissolves in it. A base for cooking knowledge would include information about measuring water or using it with different kinds of utensils. A desert-survival knowledge base would relate water to sweat and night travel. Knowledge bases for farming and boating would relay other unique information.

A partial approach to combining knowledge from these different sources is standardization. The goal of standardization is to make interchange and reuse possible. Initially, railroads were designed with different-sized gauges for different sets of tracks. Now, however, the diversity of railroad gauges has almost disappeared, and most railway cars can be routed along any set of tracks.

The same idea can be applied to create standard vocabularies (using, for example, words like *water*), standard contexts (such as *water use in cooking*), and ways of defining larger things in terms of more basic concepts. This is the conventional approach to building knowledge bases, in which the representation language for storing and reasoning with knowledge is akin to the communication language used by human experts.

Work on standardization can be coupled naturally to work on shells for expert systems. A *shell* is an environment designed to support applications of a similar nature; they are intermediate between specific applications (e.g., a system for advising engineers repairing copiers) and general purpose knowledge-engineering environments (e.g., frame-based knowledge-representation tools). Shells have been built for numerous broad applications, such as configuration tasks, scheduling tasks, and specialized office tasks. They have four crucial components besides representation languages: (1) prepackaged representations of important concepts; (2) tools for inference and representation tuned to perform the task efficiently and perspicuously; (3) specialized user interfaces for carrying out the task and entering domain knowledge; and (4) generic knowledge of the application area. For example, commercial shells for a configuration task have specialized representations of system parts and their parameters, constraints on how parts can be combined, and strategies for searching for and evaluating candidate configurations. Shells thus have the potential to be used for sharing and standardizing knowledge in communities larger than single expert-system projects. Shells are a key component in the "parts" approach to creating a knowledge medium.

Metaknowledge for Combining Knowledge

Anyone who has tried to give a computer program common sense has discovered the staggering amount of it people acquire on their way to becoming adults. And none of it is readily accessible to computers. Within the AI field many projects have attempted to model common sense reasoning and a qualitative and naive physics. Doug Lenat's CYC project (1995), which was begun in the early 1980s, is still attempting to encode the knowledge of an encyclopedia into an explicit knowledge base. Whether or not this project is successful depends on how well it deals with the additivity of knowledge.

Knowledge additivity means that when knowledge is added to a system, it is used appropriately for system tasks. *Knowledge reusability* means that knowledge added for one purpose can be reused for another. The fundamental problem is that memes are additive only when appropriate knowledge about knowledge—that is, *metaknowledge*—is available and effective in guiding the integration of new knowledge. With the right metaknowledge, adding more knowledge can become simpler as the knowledge base grows.

The shells approach to the knowledge-additivity problem is to define a context, a set of standard knowledge types, and standard ways of using the knowledge. For example, in a configuration shell the part of the system that searches for configuration solutions is capable of using the parts catalog to generate candidate solutions and of applying the standard kinds of constraints to prune the search. Knowledge for a configuration task is additive as long as it fits within the standard framework of the configuration shell. The metaknowledge in shells for knowledge combination resides in the categories, operational methods, and assumptions of use that are built into the shell architecture.

Although work with shells and experiments that store standardizing terms are the right next steps, they are only a beginning. Indeed, making standardization the only means of combining knowledge would ultimately defeat the whole enterprise.

The need for a framework embodying knowledge and intelligence to make knowledge additive and reusable has been painfully rediscovered several times. In the 1970s, many researchers in AI realized that knowledge or facts could be represented in logical steps; they believed, then; that all they needed to be sure that the knowledge would be used appropriately for any purpose was a sufficiently powerful and speedy theorem prover. They thought they could always add a few more facts and derive the consequences automatically. Knowledge additivity and reusability were not seen as problems. Just pop in the facts and turn the crank on the theorem prover. What's the problem?

The first widely recognized doubt emerged when researchers realized that theorem provers could be correct but were notoriously inefficient at solving realistic problems. Indeed, they were so inefficient that even for routine tasks a theorem prover might run for thousands of years without delivering a solution. In the 1980s AI researchers shifted their attention to developing efficient search processes and encoding knowledge to guide searches heuristically.

There was an important lesson in the experience: by itself, a theorem prover is a fundamentally ignorant system. A theorem prover depends on and benefits from standardization of terms, but it lacks the metaknowledge needed to guide knowledge additivity. Indeed, a tacit, and unfortunate, assumption of those who design theorem provers has been that all they needed to enable knowledge additivity was uniform syntax and semantics. The same flawed notion has arisen in other visions of how to build intelligent systems. For example, a similar story involves the appeal of, and ultimate disappointment in, schemes to encode knowledge in terms of production rules or in logic-based programming languages. Additivity and reusability of knowledge require more than just a database of uniformly encoded statements and a simple interpreter. Knowledge additivity in particular requires a fund of metaknowledge used by an agent who can combine knowledge.

Effective and Ineffective Committees

To put the knowledge-additivity issue in experiential terms, imagine that we try to impart knowledge to children by telling them things that they are not yet able to assimilate productively. Child psychologists tell us that until most children reach their early teens, they are unable to work fluently with abstract ideas. Even adults are often unable to make use of ideas too far removed from their familiar experience. In these cases we say that people lack the context for understanding or using the knowledge. That context is just the sort of framework that researchers design knowledge shells to provide, albeit in simplistic forms. In this respect, theorem provers know profoundly less than a young child or even the shell of an expert system. Without a context, it is not realistic to expect knowledge systems to integrate and effectively use a wide range of facts.

Let's consider another, similar example: the committees and task forces often organized to bring together the knowledge of various experts or constituencies. Most people's experience of such groups is mixed. Some of them are very productive and come up with solutions or ideas wonderfully superior to anything that individual members could have created alone. We have probably also participated in groups whose work was arguably a waste of time and yielded results far lower in quality than what any one individual in the group could have produced. What makes the difference? The key to a committee's success at combining knowledge is the metaknowledge they use to bring together and integrate what they know separately.

A dramatic example of the poor use of metaknowledge by a committee was provided by a 1996 chess match. While world chess champion Gary Kasparov was gearing up for his historical rematch with IBM's Deep Blue chess computer, grandmaster Anatoli Karpov was using the World Wide Web to take on everybody in the world. The world's chess players took on Karpov not individually but collectively. Under the sponsorship of Telecom Finland, on-line participants formed a committee to play against Karpov. Anyone with access to a computer could go to a designated web page and vote on the possible choices for each move in the game. About two hundred people voted on every move; the computer tallied their responses and picked out the most popular move of the "world chess committee." Chess experts were probably not surprised when

Karpov won handily. The committee played very conservatively, and Karpov took the offensive on the sixth move. A blunder on the sixteenth move sealed the fate of the world committee, leading to its surrender on the thirty-second move. What this example suggests about knowledge combination is that "averaging" strategies for knowledge combination lead to conservatism and mediocrity.

In this example, the metaknowledge used to combine knowledge was simply unweighted voting. What this voting failed to account for was the fact that all committee members were not equal. They had different levels of skill and, undoubtedly, different areas of chess expertise—such as openings, middle games, and endgames. If a committee consisting of an electrician, a plumber, and a carpenter were designing a house, such a voting scheme would give all their votes equal weight in deciding what kind of wiring to use.

For the past ten years Bernardo Huberman and Tad Hogg have been modeling the performances and results of cooperating agents solving hard problems (1995). Their studies provide insights into the issues of knowledge additivity, the value of diversity, and the effectiveness of groups. Their work demonstrates a universal law: that highly cooperative systems, when sufficiently large, can display universal performance characteristics not derived from either the details of members' individual processes or the particular problem being attacked. This law predicts that rather than following the familiar bell-shaped Gaussian curve of performance, such groups can perform according to a log-normal distribution—a significant difference, because the log-normal distribution has a very long tail in which the wonderful performance of effective groups resides. This prediction is expected to apply to any problem whose solution requires successful completion of a number of nearly independent steps or subtasks as long as agents performing the subtasks exchange hints about their partial results. Cooperation in the form of hints exchanged among agents leads to a nonlinear increase in overall system performance. That occurs because any collection of cooperating agents is likely to have a few high performers. In a system with *n* agents, the expected top performer will be in the top 100/*n* percentile. Huberman and Hogg's model quantifies improvements in performance according to parameters describing the search space, the total number of agents, the use of hints, and so on.

Reflecting back on the chess example, the reasons behind the poor performance of the world chess committee now become clear. Although the number of participating agents—two hundred—was significant, there was no division of labor and no communication among members of the committee. Whatever potential strength in diversity and numbers the chess committee possessed was squandered by a voting system that lacked appropriate metaknowledge for combining what the members knew. In contrast, familiar and complex examples in the real world—such as operational economies and ecologies—thrive on diversity, specialization, and communication.

Roles in a Knowledge Market: Toward a Knowledge "Services" Approach

Try to imagine an Internet-based computational knowledge market. Taking a kind of industrial systems view of it, we could ask, "What different roles does this market require?" Our knowledge of the economics model would lead us to expect knowledge producers, knowledge distributors (publishers), and knowledge consumers. Drawing on our practical knowledge of working markets we might predict many other roles. Perhaps there would be knowledge advertisers and knowledge advertising agencies. If the market were regulated, knowledge certifiers might be needed. Perhaps there would be substantial horizontal spread in our knowledge ecology, with knowledge providers functioning in many specialized niches. There might be a broad distinction between those who advise others on how to find and apply knowledge (knowledge consultants) and those who offer knowledge services that apply knowledge to some end (knowledge appliers or servers).

Experts who have participated in the creation of expert systems commonly report that the process of articulating their knowledge in order to represent it on computers has, itself, yielded a better body of knowledge and a more complete understanding of what they know. Reflecting on this experience, Donald Michie has proposed the creation of *knowledge refineries,* where such processes could be used routinely to purify and formalize crude knowledge (1983).

The model of the scientific community yields a different set of ideas about the differentiation of roles in our knowledge market. Integrators would combine knowledge from different places. Translators would move information between subfields, converting the jargon as needed. Summarizers and teachers would also be needed.

A networked economy of knowledge servers could have all the same structure as a physical economy of manufacturers of hard goods and still not address all the issues salient to creating a knowledge medium. A car can be manufactured by the coordinated work of a tire manufacturer, glass manufacturer, engine manufacturer, and body manufacturer. Similarly, it can be designed (a knowledge-intensive task) by the coordinated work of a tire designer, window designer, engine designer, and body designer. The example is oversimplified, but it illustrates the principle. A knowledge economy needs something quite different: a community of knowledge servers capable of performing knowledge work and of exchanging knowledge goods that represent, in a mutually intelligible way, the results of that work. Such communities of knowledge services already exist in the intranet supply chains of many companies. Thus there are interlinked on-line services that manage billing, inventory, order filling, and various other "back office" functions. They fall short of being knowledge media, however, because knowledge itself is not part of what the networks produce, exchange, and consume. In the present state of the art, incremental advances in the use and combination of *microtheories* seem like good roads to progress in this area.

Is it practical yet to create a knowledge medium based on computational knowledge services? The current state of the art falls short of realizing this possibility fully. Nonetheless, there are growing examples of on-line services—some of which are knowledge services. For example, there are configuration systems on-line that use an underlying shell. The role of the Net in these cases is presently limited to delivering the service to people; it plays no role in combining multiple knowledge services.

Where the Net is playing a significant role is in the delivery of documents. As more documents move on-line—on the Internet or on corporate intranets—there is an opportunity for what Xerox, IBM, and knowledge-management and document-management companies call *knowledge sharing*, in which the combining of knowledge is carried out by people.

As suggested in chapter 5, there is a real opportunity for new kinds of technology—such as sensemaking technology—to help people find and combine knowledge from multiple sources. Such endeavors illustrate the third approach to trafficking in knowledge in a knowledge medium—"documents," as contrasted with "parts" and "services."

Although the documents approach may seem ordinary and low tech when compared with the AI techniques underlying the other approaches, it is not without its issues and edges. In earlier chapters we discussed the technological challenges to creating portable document readers; considered the difficulties involved in building the trusted systems needed to provide a secure economic basis for knowledge sharing; and looked at using copyright law and digital contracts as legal foundations for a document-based knowledge medium. The Internet edge is presenting us with plenty of challenges to creating a knowledge medium, even in a rudimentary digital form.

Bootstrapping a Knowledge Medium

The foregoing discussion might make realizing our vision of a knowledge medium seem very distant. How could we bootstrap such a process? The goal of building a new knowledge medium is not to replace work on expert systems with something completely different, or to replace network-based communications media. The goal is to tie these two elements into a greater whole. A knowledge medium based on AI technology would be part of a continuum. Unlike books and other passive media, which simply store knowledge, expert systems can store and also apply knowledge. In between are a number of hybrid systems in which most of the knowledge processing is done by people. Even now, any opportunities for establishing human-machine partnerships and automating tasks incrementally already exist.

One example of a knowledge medium using AI technology is the Trillium project used at Xerox in the late 1980s. The system, designed by Austin Henderson and others, created a knowledge economy based on the memes of interface design for photocopiers. Because copiers, printers, and multi-functional devices possess many more-powerful functions than the simple early machines, the design of user interfaces has become much more challenging. From one perspective, we could view Trillium is a sort of computer-aided tool for designing the controls of a copier. It provides a language for expressing the actions of a copier interface in terms of buttons, lights, actions, and constraints. Initially, Trillium was intended to model prototypes of interface designs rapidly so that designers could quickly try them out and study the human cognitive factors related to their use. As Trillium evolved quickly into the best medium for describing copier interfaces, several other design teams at Xerox began using it. Soon these teams, wanting to exchange their design concepts, developed software to help combine design concepts, and different versions of a prototype evolved at various sites. Trillium became a major (if not the major) medium for exchanging knowledge about user interfaces. It was used to design most of the interfaces of a whole generation of Xerox copiers.

Trillium was not, however, conceived as an expert systems project but, rather, as a means of augmenting existing media—in this case, telephones and e-mail—and it did so very successfully. The main benefit of Trillium in this respect lies in its ability to express the memes of interface design, which are tangible artifacts in a knowledge medium.

Another example of a knowledge-medium project at Xerox is the Eureka project, which was developed in the late 1990s. It started with a field study of on-site diagnostics for photocopiers and highlighted the importance of field know-how. Eureka's technology consists of a database of tips and hints for diagnosing and repairing Xerox printers and copiers. It gives members of service forces the power to create and shape the working documentation of their communities, reflecting both their experience and the particular local differences in practice they have developed. Convincing and substantial productivity improvement was observed as the tips and hints in the database were contributed by some field representatives and checked by others. Eureka shows how the viability and health of an electronic medium is enhanced when it enables community members to organize and integrate their knowledge.

Toward a Knowledge Medium

We have seen that the overall goals of building a knowledge medium are quite distinct from those conventionally embraced by AI. The vision of AI, as its name suggests, is the understanding and building of an autonomous artificial intelligence. Creating such an intelligence is compatible, and probably synergistic, with shaping a knowledge medium; and the enterprise of building a knowledge medium shares much of the technology that has become important to AI. Yet breakthroughs in AI are not prerequisites for building or experimenting with such media. Intelligence developed by AI can always be added to a knowledge medium incrementally at later stages.

In 1977 Ira Goldstein and Seymour Papert announced a shift of paradigm in AI from a power-oriented theory of intelligence to a knowledge-oriented one. The fundamental problem for understanding intelligence, they argued, is not to identify a few powerful techniques but rather to learn how to represent large amounts of knowledge in a fashion that permits their effective use and interaction. Now, more than twenty years later, we recognize that the bottleneck process in building expert systems is getting knowledge into those systems (knowledge acquisition) and subsequently modifying it and updating it over time. This recognition puts the field of AI in a position to shift even closer to the foundations of knowledge: instead of focusing only on mechanisms of intelligence, AI is now studying the role of knowledge in intelligent systems and looking for methods to augment the knowledge processes of a medium.

The conventional, older goal of AI leads to projects for which the creators can say, "Look ma, no hands!" The knowledge medium, however, requires us to change our goal from product to process and to introduce new criteria for success. There are important questions to be asked. As a project evolves, where will the new knowledge come from? What experiences will drive the creation of new knowledge? How will it be distributed and who will use it? What kinds of knowledge will be distributed and what form will this knowledge take?

Reflections

In the late 1980s, many people in the AI and computer business world were watching the formation of the Japanese Fifth Generation Project. The original proposal for this scheme described a number of roles for knowledge-processing systems: increasing their intelligence to better assist humanity, putting stored knowledge to practical use, learning, and associating data. In his keynote speech for the second ICOT conference, Hiroo Kinoshita of the Japanese Ministry of International Trade and Industry hailed the creation of an advanced information society.

> In it, different information systems will be linked into networks, and a variety of services will be offered. In addition, rather than individuals playing the passive role

of merely receiving information, they will be able to obtain that information which they require, use it, and transmit it among themselves, in what is expected to be a society more closely reflecting human nature. (Kinoshita 1985)

This goal proved, ultimately, too ambitious for the Fifth Generation Project, which subsequently focused on narrower objectives and abandoned its plans to build experimental knowledge ecologies or knowledge markets. It did, however, demonstrate the process of planning for an earlier journey to the Internet edge. The network Kinoshita describes would be a knowledge medium in the sense we have used this term. Kinoshita, a policymaker, mentioned several of the practical difficulties of bringing the network into existence: the lengthy period needed to write software and develop mechanisms (and knowledge) for computer security, interoperability, and better human-machine interfaces. His concerns are still valid and focus on some of the technical challenges we need to create a vibrant knowledge medium.

The effect on the human condition of creating a knowledge medium is potentially tremendous. Such a medium could greatly enhance our ability as a culture to create and use knowledge. At the same time, the technology for knowledge media could provoke further pushbacks from the Internet edge. By augmenting the change amplifier described in chapter 1, automatic knowledge processes might well upset existing knowledge equilibria—creating new and unexpected knowledge and insights—as memes cross the until-now-impermeable boundaries of human specialization.

The next knowledge medium will be a variant of our information market dream that applies greater automation to the creation and use of knowledge. In the 1980s and 1990s, elements of this dream appeared in the form of data-mining technology and communication support for people in a networked medium. With e-mail and the technologies of digital publishing, the Net already functions as a communications medium. Before it can also function in a significant way as a knowledge medium, however, we will need to address the difficult issues of knowledge additivity and reusability.

7

The Edge of Chaos: Coping with Rapid Change

Complex systems seem to strike a balance between the need for order and the imperative to change. . . . [They] locate at a place we call "the edge of chaos" . . . where there is enough innovation to keep a living system vibrant, and enough stability to keep it from collapsing into anarchy.

Michael Crichton, *The Lost World*

Since the beginning of the decade, the struggling interactive media industry seems to have been racking up bad karma by forcing itself into existence before its time. Its ongoing delusion [is] that its business is the Next Big Consumer Trend . . . first on CD-ROM, and now on the Internet.

Of course, any time that kind of investment money starts flying around, everyone gets very excited. . . . But now the dust has begun to settle and the grim reality has set in: customers are not interested. Money is not being made. . . . Something has to change.

Denise Caruso, "Digital Commerce"

Both the beginning and the closing years of the twentieth century were characterized by intense innovation and change. Between 1890 and 1920, automobiles, electric lights, telephones, radios, movies, phonographs, and numerous other inventions entered the everyday lives of people in the developed world. The last decades of the century also introduced many inventions—personal computers, microwave ovens, cellular telephones, compact discs, VCRs, and countless innovations in electronics and optics to power new forms of communication like the Internet. The computer industry came to expect rapid improvement and early obsolescence of its systems, while consumers and organizations struggled to balance the potential benefits of acquiring new technology against the predictability of retaining older technologies in an installed base. The accelerating pace of life itself suggested that some variation of Moore's Law (1991)—which holds that computer processing speed increases 44 percent per year—was at work. In 1997 the phrase *Internet years* came into vogue to describe the phenomenon that innovations in technologies and businesses related to the Net seemed to multiply several times faster than those off the Net.

In chapter 1 I portray the Net as a change amplifier. In the present chapter we consider the process of change itself, focusing on the rise and fall of technological innovations and their associated businesses. Historically, change begets more change and, when it compounds very quickly, begets chaos. The resulting social effects lead to pushbacks—unwillingness to adopt new technologies or refinements to older systems.

To examine change as a phenomenon, we consider work at the technological edge and at the different stages of Geoffrey Moore's model of the technology adoption cycle. To do so, we draw on two kinds of historical sources. Because their experiences are so crucial to the continuing evolution of the Net, we consider stories about some early travelers to the Internet edge. In addition, to provide depth and perspective on the technology cycle, we look at several stories about the Erie Canal. Scattered throughout the chapter, they can help us see phenomena present at the Internet edge but obscured by the freshness of our Net experiences.

Stages of Technology Adoption

When a change in how some element of one's business becomes an order of magnitude larger than what that business is accustomed to, then all bets are off. There's wind, and then there's a typhoon.

Andrew Grove, *Only the Paranoid Survive*

During periods of rapid social and technological evolution, changes are felt by different people at different times. In his books *Crossing the Chasm* (1991) and *Inside the Tornado* (1995), Geoffrey Moore draws a bell-shaped curve to illustrate what he calls the *technology adoption life cycle*. His curve, depicted in figure 7.1, organizes the cycle into stages and names the players or roles pertinent to the various stages of a technology's development and marketing. Each stage requires specific kinds of edge work as players confront the particular issues related to change at that stage. Together, the stages describe the process of a social group meeting a collective edge as the influence of a technology rises and falls.

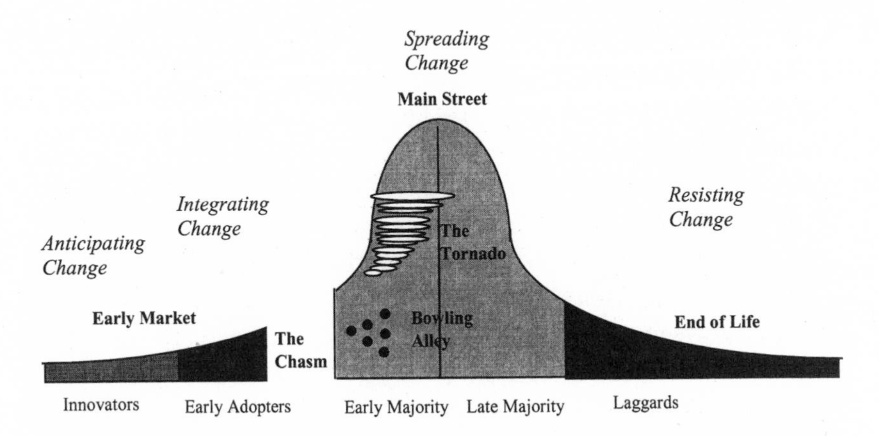

Figure 7.1
Technology Adoption Life Cycle. Changes are felt by different people at different times. Geoffrey Moore describes the staged process of technology adoption in which the influence of a technology rises and falls.

The first stage, the early market, is a time of great excitement when technology enthusiasts are eager to try the new technology. Moore characterizes the players in this stage as *innovators* and *early adopters* and includes a third, external actor, the *technology watcher.* The innovators pursue products aggressively, sometimes even before a formal marketing program has been launched. Their interest focuses on pilot projects to test the applicability of the technology. Next, early adopters—who should perhaps be called *early adapters*—are willing to base buying decisions on a new technology that is a potentially good match to their needs. Much of what they do involves making adjustments to integrate the new technology into existing but unprepared work settings and practices. The technology watchers are active during this stage and before, reporting on emerging technologies and the pilot projects conducted by the innovators. These three actors—the innovators, early adopters, and technology watchers—are the first to journey to the edge with a new technology.

Moore's principal interest, however, is in technologies that make it big and in business strategies for company growth. He recognizes that the integration work innovators and early adopters must do to use the new technology is much greater than less-motivated consumers in a big market will accept. That integration work brings us to the stage Moore calls the *chasm,* when early market enthusiasm has waned but the technology is still too immature for mass adoption. Many technologies stall at this point and are eventually overtaken by other approaches. Successful promoters at a chasm look for applications that can craft the new technology to fit the needs of multiple niche markets. Another of Moore's metaphors for this period, the *bowling alley,* likens promoters' efforts to aim a technology at particular markets to targeting a bowling ball at certain pins. If a company chooses its niches wisely, it gains other niches as well (knocks over other pins) and travels up the learning curve with a refined technology that meets the needs of a larger market.

If a technology successfully crosses the chasm, what follows is a period of mass-market adoption when many people—the *early majority*—switch over to the maturing technology. Because of the chaos and rapid growth that accompany it, Moore calls this stage the *tornado.* The rapid growth quickly moves the technology, and consequent social and work changes, through a large population of users. When a technology rides the tornado it reaches the next stage, which Moore refers to as Main Street. At Main Street the technology is picked up by the *late majority,* people who prefer to wait until a technology has become an established standard and is usually less expensive.

Even after a technology is widely deployed and becomes the new installed base, it can stagnate. If there is too little turnover and evolution of the technology, it can become the target of newer, competing technologies. At this stage, the mass of users and the market inertia created by that large installed base can inhibit necessary changes. To extend their stay on Main Street, producers of an evolving technology and the businesses based on it have to defend the base against newer technologies. They can do so by leveraging their market presence and income to finance continuous, incremental improvements that stave off displacement by next-wave technologies. However, *backward compatibility*—refinements to a product that is still usable on systems produced earlier—can be both an asset and a liability, as we will see below. Backward compatibility takes advantage of the installed base but, at the same time, puts limits on the degree to which a technology can change. A technology provider must determine when the need for incompatible changes outweighs the benefits of compatibility and needs to have a transition strategy to avoid losing old customers.

Anticipating Change

Although I have an ardent wish to live and see many of them effected, yet, by accident, I may be writing for a subsequent age: And I have that reliance on the American character, already established for its inventive genius and enterprise, which gives me even grateful expectations that . . . my countrymen are capable of encountering many difficulties and apparent impossibilities, by which many improvements . . . will be undertaken and completed in a future day. . . . it would be a burlesque on civilization and the useful arts, for the inventive and enterprising genius of European Americans, with their large bodies and streams of fresh water for inland navigation to be contented with navigating farm brooks in bark canoes.

Jesse Hawley, 1807

Depend upon it. If we vote for the canal this day, we mortgage the State forever.

Representative Emmot, a legislator from eastern New York, 1815

Even before a technology comes into wide use, dreamers and technology watchers may write about it. Jesse Hawley was such a dreamer in nineteenth-century New York. When he found himself in debtor's prison for twenty months he began writing a series of fourteen essays describing the benefits to be obtained by building an overland canal linking the Hudson River with Lake Erie. Hawley's essays, signed "Hercules," were published in the Genessee (N.Y.) Messenger in 1807 and 1808. Thus Hawley the dreamer became the first publicist for the Erie Canal, which was completed in 1825.

Not all dreamers and technology watchers are promoters. Some are more aware of its dark side—either unwanted social effects or undesirable expenses. The quotation from the state legislator also dates from the period before work on the canal began. Emmot, realizing that the Erie Canal would be a very large and expensive public works project, objected to the special tax assessments proposed for cities along the route.

Technology watchers such as Hawley and Emmot spoke to a public trying to sort through the predictions and chaos to understand the choices ahead. In the following passages, we revisit the similar chaos experienced by early–1980s technology watchers as they considered several promising technologies for electronic publishing.

Watching Technology: Pushbacks from the Internet Edge

In the United States the leading professional organization of the book publishing industry is the Association of American Publishers (AAP). In 1980, the AAP added a "Technology Watch" column to its monthly newsletter to monitor and assess emerging technologies relevant to electronic publishing (Risher 1980–1985). For publishers, electronic publishing is either a dream or a reason for dread, depending on whether they see it as a new business opportunity or a threat to existing markets. These competing views were reflected in the second column of "Technology Watch," which carried the subtitle "Electronic Publishing: Opportunity or Threat?" Absent from the title was a third, cynical, and possibly majority, view: "Bust!" In the following pages, as an example of how trend spotters work in a period of rapid change, we consider stories from the column that appeared over a four-year period.

In its first appearance in February 1980, "Technology Watch" reported on three new technologies. The first, developed by Microsonics Corporation, was a way of encoding sound in printed messages. The technology was being considered by *Time* magazine's editors to permit readers to hear the voices of people quoted in its articles. The second technology was being tested by the Xerox Corporation to modify copiers to locate copyright bar codes on documents. These copyright-sensitive copiers would produce hard copies only after recording the item in a copyright report. The third technology, two devices for reading printed text, was introduced by Kurzweil Computer Products. One version of the device converted the scanned text to synthetic speech; the other converted it to digital text.

The three technologies led to different pushbacks from the edge. The text-embedded sound messages never achieved broad use, perhaps because they required subscribers to buy handheld scanners for this purpose alone. The copyright-sensitive copier was immediately shelved when initial tests showed that to be profitable the new equipment required a volume of copying beyond what was done in most libraries and information systems. Several related devices were tested by Xerox over the next few years, but, ultimately, pushbacks came from both publishers and the competition.

Xerox determined that copiers with copy-prevention features would not sell as well as those that make unimpeded copies; and publishers calculated that expected revenues from the transactions would not cover the costs of accounting and assigning bar codes to texts. For a while, the Kurzweil products were the only viable optical character-recognition devices in the niche market. Over the next few years, however, companies with greater dedicated resources entered and expanded the market, reducing Kurzweil's market share.

Of the three technologies in this example, two failed in the market for which they were intended and the third succeeded at first but was later displaced by a competing technology. The first two failed to make the transition from innovators to early adopters, and the third failed to cross the chasm of the adoption cycle to Main Street. Such turmoil in the introduction of a technology is not unusual. When a new technology requires or enables wide-ranging changes in how something is done—as in the shift from paper publishing to electronic publishing—the potential for pushback and chaos is even greater. One way to look at this chaos is as a reflection of the interaction between creative energies and complex situations. While some promoters and early adopters are searching for things that work, others are resisting things that don't work for them for reasons promoters do not yet understand. Creative energy tries to find a way through a complex collective edge.

Four Years on Fast Forward

From the perspective of the late 1990s, early 1980s technology seems primitive and tame. Nonetheless, technology watchers trying to anticipate its effects on publishing were seeing early versions of systems and witnessing controversial pushbacks that parallel those of the 1990s. New technologies were appearing every few months, and publishers were trying to work out strategies for dealing with them. To compound the confusion, the mix of technologies offered suggested many different possible futures. To recreate the sense of anticipation and bewilderment that reigned, we trace some of the new technologies of the period as they were evaluated in "Technology Watch" over a four-year period.

1980

The opening year of *"Technology Watch"* saw the public introduction in the United States of videotex, teletext, videodiscs, and information utilities for personal computers. Each of these technologies seemed to offer new possibilities for electronic publishing. Although they all promised to deliver information to the home or office, their strategies were quite different. In 1980, technology watchers bet on whether information would be delivered on prerecorded discs or over the wire; if over the wire, they wagered on whether information would be sent to television sets, mainframes with remote computer terminals, or personal computers.

The videotex technology linked databases to television sets through telephone and cable television lines. Test users received a modified television set and a keypad. By pushing appropriate buttons, they could select from a menu of program categories including news, shopping, education, health, travel, automotive, puzzles, and entertainment. They could order advertisers' merchandise and select the stories they wanted to watch. Promoters expected videotex to grow to reach a mass audience.

In March of 1980, "Technology Watch" reported that Knight-Ridder's subsidiary, Viewdata Corporation, was wiring around two hundred homes in Coral Gables, Florida, for a videotex service called Viewtron. Publishers such as Addison-Wesley and Macmillan developed materials for the test, and the booksellers B. Dalton offered automatic ordering of books from a list of best-sellers.

One of the limiting factors in the experiments was that only a small number of television sets were suitably modified to work with videotex. To test a system, developers typically wired only fifty or sixty houses at a time, then removed the modified sets to try the system with a new batch of users.

Another technology, teletext, also used television signals to carry information. It was simpler than videotex, however, because communication was one-way. Users could select what they wanted to view, but they could not send signals back to the broadcaster. The system uses the vertical blanking interval of the standard television signal to carry information and provides users with a decoder to retrieve and view the text and graphic content embedded in the signal.

The videodiscs of 1980 were large-format discs for storing television signals. The May 1980 "Technology Watch" reported that technological advances made it possible to store up to fifty-four thousand pages of digital information on a videodisc for random-access retrieval. The possibility of using a videodisc with a computer instead of a television set was seen as opening up many new possibilities: video demonstrations at trade shows, educational applications, and storage of large databases. Different and incompatible formats of the disc were marketed by MCA/Philips, RCA, and JVC. "Technology Watch" noted that the lack of a uniform standard caused confusion for consumers and was inhibiting publishers from getting involved.

Another emerging technology concept in 1980 was the digital information utility. This technology differed from videotex and teletext in that it was based on computers instead of television sets. A user with a terminal could dial into a time-shared mainframe computer to gain access to on-line information. Some companies offered experimental services with specific databases; AT&T experimented with electronic Yellow Pages, Dun and Bradstreet tried an on-line airline guide, and Nexis and Mead Data Central provided dial-up access to stories in the *Washington Post, Newsweek,* and *The Economist.* An on-line service called the Source offered several different kinds of service. Rather than billing for access to particular information, the Source and other dial-up services based their charges on the time users were connected to the mainframe.

In December 1980, "Technology Watch" reported that several technologies were experimenting with on-line education. The Source was distributing lecture notes and offering a degree program, and videotex services were trying two-way video classes. The same issue announced a new National Science Foundation grant program for developing educational programs using personal computers. Each of the technologies introduced that year was leveraging its unique strong points: a remote video classroom experience for videotex, inexpensive and coordinated information distribution for an information utility, and individualized programmable interactivity for personal computers.

The December issue also included a discussion of potential copyright problems relating to videotex. Polaroid was promoting a CU–5 close-up camera as ideal for making photos from a television screen—which would have been perfect for making hard copies of videotex pages. Although the volume of information being distributed in videotex was not yet commercially significant, publishers were already anticipating copyright problems.

1981

In the second year of "Technology Watch" the four technologies described above continued to compete for ways to deliver information electronically. New technologies were entering the market and would soon affect the competition. Potential consumers of videodiscs for movies responded enthusiastically to a cheaper alternative as imported videocassette recorders (VCRs) entered the market in quantity. A fax machine cheap enough for consumers was introduced and considered applicable to the home and office delivery of hard copies. It could transmit a page through the telephone network in under two minutes. The same year, the first relatively inexpensive laptop computers—the TRS–80s—were introduced by Radio Shack, and Random House started developing mathematics curriculum software for them.

The February 1981 issue described several continuing experiments in delivering information electronically. Eleven AAP publishers were working together to provide a thousand "pages" for a teletext experiment in Fort Lauderdale. The information was broadcast in an electronic magazine of 125 pages; some pages were updated every half an hour, some every hour, some only daily or weekly. "Technology Watch" cited the most frequent complaint as the delay time between request for a page and its appearance on screen. Although the delay could be as little as two seconds, the average was thirty seconds.

Precursors to the Minitel system began to appear in France, where the telephone company was installing fifteen hundred test terminals in the Île et Vilaine region. The expense was justified as an alternative to printing white and yellow pages telephone directories. Approximately 270,000 more terminals were scheduled to be installed in the region in early 1982 at no cost to users. The Source Telecomputing Corporation reportedly ordered 250,000 American versions of the terminal.

The AAP saw the involvement of the French telephone company in videotex technology as a potential threat to publishing. They were afraid that telephone companies would retain monopoly control over telephone lines, dominating electronic delivery of information to the home and perhaps competing unfairly with other news sources. By August, AT&T was apparently feeling some of this pushback. "Technology Watch" reported that because of protracted regulatory and legal proceedings, AT&T had decided to cancel its planned Austin, Texas, experiment in providing electronic yellow pages.

An important part of the evolving story of the different technologies was a 1981 update on the total numbers of electronic systems deployed. By 1979, "Technology Watch" reported, 535,000 desktop computers had been installed in the United States; the column's editors estimated that there were 3.1 million by the end 1980. During the same period cable television was installed in eight million homes and was potentially available to fifty-five million more. Teletext was still in limited distribution in the United States, although there were over eleven million subscribers in the United Kingdom. According to these numbers, television-based systems had the greatest installed base worldwide and were the leading vehicle for delivery of electronic text information—in spite of their limited success in the United States. Videodiscs, on the other hand, were faring poorly and facing increasing competition from VCRs. In 1980, the number of videocassette recorders imported into the United States jumped to 920,000 units, compared with 599,000 in 1979. The 1981 imports were projected to be 1.3 million units.

From a global perspective, the picture of competing information technologies was confusing, because the outcome was different in different countries. In Sweden, teletext experiments had been conducted since 1975, and several thousand families had televisions sets with decoders. Videotex trials had begun there in 1979 but were in a very early stage. In the United Kingdom, teletext had 180,000 users—more than anywhere else in the world—receiving both public and commercial services. Videotex had only 10,877 subscribers even though it was available to 62 percent of the country. In Germany, teletext subscriptions were very low and videotex, with only 6,000 subscribers, was only beginning. There was a good deal of discussion about home delivery of information via satellite transmission.

By the end of 1981, larger trials of videotex were being developed for rural areas of the United States. In addition, developers were experimenting with charges for everything from books to groceries to determine what services consumers would pay for. On some trials users paid according to the services they chose. On others, experimenters charged a flat rate, in order to understand how consumer choices would differ if there was no disincentive to extensive use.

1982

In 1982 the stories in "Technology Watch" focused increasingly on personal computers. The industry reported that there were 150,000 personal computers in education, 200,000 in small businesses, and 300,000 for scientific use. Technology watchers predicted that there would be a computer in every home by 2001.

Publishers of software for PCs were discovering that their programs were being copied and passed among consumers without payment. The problem was rampant: it was estimated that four out of five copies of software then in use were pirated. Milliken Publishing reported that one school purchased a copy of their educational software and reproduced enough copies to supply all the schools in its district.

Publishers tried various approaches to prevent copying. They used diskettes with features allowing them to be copied but only a very limited number of times (typically two); they offered software that could not be used without a plug-in microchip; they sold systems that were unusable without printed workbooks. Texas Instruments offered a cartridge for software distribution that prevented copying, but it was only usable on their own computers. Computer networking, glimpsed on the horizon, offered yet another threat to copyright from people who could download software from the network instead of buying their own copies.

"Technology Watch" continued its world survey of electronic publishing technologies. It was a confused picture. Spain was developing an interest in videotex, Japan was focusing on VCRs and videodiscs, and most Canadian developers preferred systems based on personal computers. Australia had only one publisher bringing out electronic materials; it offered legal information on using an information utility.

The Library of Congress began a project to preserve its holdings on digital optical discs, projecting that 95,000 pages could be stored on two sides of an optical disc. The public meeting called to discuss the project raised many questions about copyright requirements for books stored on discs.

The Source announced the on-line publication of *The Blind Pharaoh* by Burke Campbell (1982). The author received royalties whenever someone read or printed the book from a terminal. The average cost to the reader on the Source was $2.80.

1983

"Technology Watch" witnessed a rise in the prospects for home computers and a loss of confidence in videotex and teletext in 1983. Small computers were reportedly in use in 48 percent of all U.S. schools. Although only 3.8 percent of all households had personal computers, a survey of credit card holders indicated that over 24 percent of families planned to buy them. Research on advanced computing was also given a boost. In the United Kingdom $300 million was being spent on new projects in computer science and artificial intelligence. In Japan, a large project, the Fifth-Generation Computing Project, was started. Senator Edward Kennedy gave an impassioned speech before Congress urging federal support for similar U.S. research—including work on an infrastructure for networking the universities. "Technology Watch" noted the effects of Moore's Law in its account of a 256K-bit chip manufactured by AT&T. It also reported enthusiastically on the rise and benefits of electronic mail.

Difficulties with videotex were now being voiced more openly. Although it was featured at a big technology show in New York, the column's editors noticed no particularly worthwhile system innovations. The slow pace at which information was displayed had still not been addressed, making videotex unsuitable for all but the briefest displays (such as weather updates). "Technology Watch" speculated that videotex and teletext might soon be overcome by competing technology.

In January, the Library of Congress announced it was studying the use of optical discs for storing photographs. They contracted for a project that used the same technology as the Adonis consortium, a group of publishers that included Blackwell Scientific Publications, Elsevier Science Publishers, and Springer-Verlag. By March, however, the consortium was rethinking its commitment to optical discs.

Futurists were already anticipating the portable document reader (PDR) discussed in chapter 2. Addressing a Subcommittee of Congress looking into issues of technology and copyright law, Benjamin Compaine, executive director of Harvard University's Program on Information Resources Policy, said that the terms used to describe media limited our understanding of the future. He suggested that we think of media in terms of content (the information), process (how it was distributed), and format (how it was presented). He noted that the lines of demarcation between various types of media are increasingly blurred; for example, that book and newspaper publishers were delivering text over television, cable, and computer terminals. He predicted that newspapers would be delivered to subscribers electronically on portable, flat, high-resolution screens.

Among other technological developments, voicemail was introduced and Grolier's began offering an on-line encyclopedia to schools in New Jersey.

How the Stories Turned Out

Whenever we hear the first part of a tangled and engaging story, we wonder how it turned out. Although the themes in the stories that began in 1980 continue, in the interim some of the subplots have been resolved.

By 1986, fewer than five thousand homes in the United States had teletext viewers, encouraging broadcasters to explore other uses for the blank spaces in the television signal. The Viewtron system built by Knight-Ridder in Florida closed in 1986 after losing over $50 million. On the West Coast in Orange County, California, the Gateway videotex system built by *Times-Mirror* Publishing met a similar fate. In terms of the technology adoption cycle, videotex and teletext never made it across the chasm in the United States, although they are still widely used in Europe. As I write in the late 1990s, two-thirds of all television sets in the United Kingdom can receive teletext and surveys show that about nineteen million people a week use versions of the technology offered by Teletext Ltd.

Why did television-based systems do so much better in Europe, even with arguably less-advanced technology? Several factors probably played a role. In the 1970s, the British Broadcasting Company still dominated television broadcasting in the U.K.; its decision to use teletext meant that set makers had to provide the capability. In Great Britain, France, and Germany, government subsidies helped establish the communications infrastructure for teletext. In the United States the federal government was then in the process of breaking apart AT&T in the name of the free market. In that context, government subsidy for a communications infrastructure was virtually unthinkable. Without such subsidies, no U.S. company could risk establishing the infrastructure needed in such a large country. Furthermore, commercially dominated U.S. broadcasters worried that a teletex service would compete with their regular programming.

At the end of the 1990s, however, teletext in Europe was showing its age. Compared with the personal computer, its graphics are primitive, and only a limited number of characters can be displayed at once. Nonetheless, users do not have to buy a computer, can access teletext information via their set's remote control, and pay no monthly fee.

The development road for digital information services has been quite different, but equally rocky. The Source was purchased, first by *Reader's Digest* and later by CompuServe. In 1997, CompuServe itself was purchased by America Online, which had bested its competitors in the crucial tornado stage of the technology adoption cycle. Videodiscs faded from the scene. The expense of laser discs, as well as the consumers' inability to make copies, meant the technology could not compete with VCRs. Videodiscs and optical discs for storing digital information were also displaced by the CD-ROM, the computer variant of the compact disc (CD) for music. In the late 1990s, CD-ROMs, in turn, were being challenged by the higher-density digital video discs (DVDs).

Since the early 1980s the contest between television sets and computers as vehicles for information delivery has shifted away from competition and toward convergence. One reflection of that convergence is WebTV and other systems using set-top boxes. Another is high-definition digital television. In 1996 more computers than television sets were sold in the United States, though that is not an absolute measure of popularity. Approximately 40 percent of U.S. homes had computers in 1997, whereas essentially 100 percent had television. The sales figures remind us that computers become obsolete and are replaced more frequently than television sets. This rapid replacement cycle indicates that the computer industry and related technologies are evolving more rapidly. If that trend continues, computers may define new niches while television, by comparison, stands still.

In chapter 2 I suggested that PDRs will be better platforms for reading and information retrieval than either television sets or desktop computers. In 1985, "Technology Watch" reported that a company was soliciting publisher investments to build a PDR or "electronic book" based on a microprocessor, liquid crystal display, and diskettes for digital book delivery. At the time, the technology constraints for building PDRs were much more severe than those we considered in chapter 2. The dim and low-contrast LCD displays available in 1985, in particular, were much less suitable for reading. Furthermore, in the mid–1980s publishers saw little advantage in investing in an unproven technology that had the potential to cannibalize their business instead of expanding it. Publishers focused their business strategies on remaining content providers and copyright holders and decided to let others take the risk of creating and popularizing a PDR technology; publishers would then license the copyrighted works to these companies. As the new millennium begins, a reassessment of the practicability of PDRs and their impact on digital publishing lies ahead.

The Role of Forecasting

In the early stages of the technology adoption cycle technology watchers and trend spotters are among the first to react to new technologies. Their job is to spot interesting technologies, study new products and the innovators using them, evaluate their potential, and pass the news along to consumers and promoters. At this stage, when the full potential of a new technology has not yet been explored and the likely pushbacks it may evoke are still unknown, watchers' assessments can easily overestimate or underestimate its value.

One of the ways technology promoters and investors manage risk is by developing small-scale pilot projects to sharpen those assessments. "Technology Watch" reported on several projects of this kind for videotex and teletext—including those in Coral Gables and Fort Lauderdale. These projects, intended to determine what information services people would use and what they would be willing to pay for them, revealed problems with the speed of information delivery. The problems were noticed early but apparently not addressed. Compared with the Net, these projects, which involved only hundreds of people and only hundreds of pages of information, seem minuscule. It is sobering to note, however, that even though in the late 1990s the Net was millions of times larger than these projects, the question of delivery speed was still relevant.

With hindsight, it is easy to be smug about the false starts of old technology, but it is more productive to find lessons in the earlier efforts. Several decades of attempts to create an information marketplace suggest that the Internet edge for this dream is not simple. We aren't through it yet, and each generation must face the challenges.

Integrating Change

In the jargon of Silicon Valley, technology watchers "talk the talk" but don't "walk the walk." They see and report. They are opinion leaders and can increase the attention paid to a new technology or sink it early by debunking or ignoring it. However, they are not directly involved in making new technology work. Innovators and early adopters, however, do walk on the front line toward the edge. For them walking the walk can present profound challenges related to integrating a new technology within the larger systems of companies and economies.

Erie Canal Stories—Bootstrapping and Using What's at Hand

> In 1818, after repeated experiments with varieties of limestone, [Canvass White, a young engineer] found a type in Madison County that when made into a quicklime cement, had the particular virtue of hardening under water. Once under water, the cement became increasingly hard with age. Abundantly available and easily prepared, this limestone produced a cement superior to any found in America. . . . The discovery and application of this cement in New York is one of the epochal achievements in the building of the Erie Canal.
>
> Ronald Shaw, *A History of the Erie Canal*

Engineers proposed to run the Erie Canal 350 miles through northern New York from Albany to Buffalo. Along the way the terrain varied widely, and so did the engineering challenges. The middle section—the ninety-four-mile stretch between the Mohawk and Seneca rivers—was the easiest part of the route and became the proving ground. Work in the west and east came later.

In 1815, no American engineers were expert enough in canal building to train others. The canal itself became a school of engineering. Young men joined the canal in many capacities and were apprenticed to more experienced men to learn the elements of surveying, design, and construction. Workers moved forward as the canal grew in length, gaining experience with each new section. In the introduction to his 1821 book of documents on the canal, Charles G. Haines boasted about this on-the-job training: "For accuracy, despatch and science, we can now present a corps of engineers equal to any in the world. . . . The canal line is now one of the most excellent schools that could be divised [sic], to accomplish men for this pursuit" (quoted in Walker and Walker 1963).

Not only was there an initial shortage of engineers, there was a shortage of knowledge about working efficiently in the wilderness. Highways and trucks to bring distant materials to the canal site were, of course, nonexistent.

White's discovery of a cement that could harden underwater was a pivotal event that greatly simplified the process of lining the canal to prevent seepage. Competitive bidding led to other ingenious devices as contractors vied to make a profit from canal building. Some inventions of a rather mundane sort made great practical differences. One worker, Jeremiah Brainard of Rome (N.Y.) invented a new wheelbarrow that was lighter, more durable, and easier to unload than previous models; its bottom and sides were made of a single board bent to a semicircular shape. Contractors soon learned that a plow and scraper to mix cement was more efficient than the European method of using a spade and wheelbarrow. A cable secured up the trunk of a tree could be pulled by turning a crank on an endless screw and a roller. This enabled a man to fell a tree without using an ax or saw. All these inventions contributed to managing the challenge of wilderness canal building.

In short, building the Erie Canal spawned more than just one new technology. Its success lay in developing new devices and methods as they were needed and training people to use them. Such creative discovery and training was a fundamental part of the challenge of integration presented by the canal.

Upgrading Technology—A Librarian's Perspective

Change comes not in a whole deal but in fits and spurts. You see something that could be useful to you and you want to introduce it, but it doesn't come as part of an integrated system.
Giuliana Lavendel, 1997

Giuliana Lavendel, head of the Information Center at Xerox PARC, is a good example of an early adopter. Libraries are always under pressure to have the latest technology and, at the same time, to integrate the technologies already in place. As Lavendel puts it, that's like trying to build on quicksand. The information manager must be extremely flexible and agile, keeping current with new technologies while working within a constrained budget.

The CD-ROM is generally considered a successful technology that rode the tornado to rapid deployment. By the mid–1990s, Grolier's had sold over a million copies of its CD-ROM encyclopedia, and most personal computers came equipped with drives for CD-ROMs.

Each CD-ROM disc can store about 650 megabytes of information. But how much storage capacity is enough? The answer depends on your point of view. One of the goals of CD-ROM development was to make large collections of information available on-line, which makes libraries a target market for the technology. In the mid–1990s Lavendel wrestled with the question of whether to buy CD-ROM towers for her customers—information consumers at Xerox PARC in Palo Alto and many other Xerox locations. She noted that in the mid–1990s the complete publications of the Institute of Electrical and Electronic Engineers (IEEE) would occupy 270 CD-ROMs. She concluded that even with towers that hold multiple drives, the logistics of maintaining an on-line collection were formidable.

In the late–1990s, as rival DVD companies promoted divergent standards, the issue of media capacity remained unresolved. Even with adoption of the proposed two-layer, double-sided DVD with a capacity of seventeen gigabytes—twenty-six times the capacity of a CD-ROM—it would still take about ten DVD players to hold the IEEE collection—and that is just the output of one publisher. Nor should we expect annual incremental improvements in DVD capacity. The whole approach to establishing standard formats for publishing media assumes that there is a large installed base of players for the media. We cannot increase density every year without also upgrading the players. Suppose, for example, that each year developers introduced higher-density media for use with backward-compatible players. Publishers, however, would probably resist publishing on the new higher-density DVDs, because they could not reach people who owned only older players. In a large organization that upgrades some of its equipment each year, users would face a guessing game to determine which players are compatible with which discs. Investments in media formats therefore tend to be generational. Once a format has been widely adopted, a generation lasts five to ten years or more.

At about the time that Lavendel was facing the question of investing in CD-ROM towers, the World Wide Web began its period of rapid growth on the Internet. In principle, libraries could access electronic copies over the Net—switching the function of storage and media access back to publishers. However, given libraries' historical role as repositories of information, having access to copies at the publishers through the Internet is, though practical, a more radical suggestion than investing in CD-ROM towers. Moreover, Net access has its own challenges, such as how to protect digital rights (see chapter 3) and how to ensure that response times will be rapid enough to let browsing readers follow links in a hypertext corpus. Depending on the Net for access also troubles librarians, because if they do not store journals on their own premises they cannot be sure of having continued access to back issues if they decide to cancel a subscription. Major debates over how to maintain the functionality and traditional fair use policies of paper publishing with on-line access continue.

Not all early adopters, of course, made the same decision about CD-ROM towers that Lavendel did. She points to colleagues in other fields, like the aerospace industry, who invested in a massive change: they bought large, turnkey CD-ROM tower systems. A year after installation, the systems were already becoming obsolete. By contrast, Lavendel describes her strategy as an information manager as year-at-a-time budgeting: "You don't try to change everything. You change this part, and then that part."

Sometimes a piece of technology lasts a long time in an organization, even after it is quite outdated. Lavendel recalls that the PARC Information Center once had an old Texas Instruments computer and printer used for a specialized task. After many years, someone proposed replacing it with more modern equipment. A librarian who did not want to switch to new equipment vigorously opposed the change and the machine remained. This situation went on for several years until a student intern accidentally spilled a cup of coffee on the machine and it died.

The advocate for keeping the old technology wanted to avoid the time and costs involved in retraining, an argument that was not limited to conservative "diehards." People who measure increases in worker productivity often point out that knowledge workers generally do not achieve greater productivity with information technology. By their measures, information technology (IT) compares unfavorably with such innovations as the telegraph, which speeded communication a thousand times, or the typewriter, which was three times as fast as handwriting. A serious aspect of the productivity problem for IT is that its rapid evolutionary cycle leaves users in a perpetual learning mode.

Spreading Change

> If the IT community moves too soon, they incur all the trials of early—which is to say premature—adoption, devoting precious resources to debugging systems . . . and stretching themselves thin running systems in parallel until the new paradigm is reliable. . . . If they move too late, on the other hand, they expose their company to competitive disadvantages as others in their industry operate at lower cost and greater speed by virtue of their more efficient infrastructures. Worst of all, if they move way too late . . . they run the risk of getting trapped in end-of-life systems.
>
> Geoffrey Moore, *Inside the Tornado*

For a new technology to grow past niche markets, it must have a broad value and appeal. For example, it might enable people to do something they could not do before, or do something much better or faster than they could before. That value and appeal is the force that spins the tornado during the period of rapid growth and spreading change.

In his analysis of the tornado phenomenon in the information technology industry, Moore looks at the support groups that cut across company and industry boundaries (1995). As he sees it, these groups are united to answer a single question: Is it time to move yet? Executives find themselves balancing the risks of moving too soon with the risks of moving too late. As the time to shift approaches, the two risks become about equal. This, he notes, creates an instability.

As a way of coping they reach agreement about three principles:

1. When it is time to move, let us all move together. . . .
2. When we pick the vendor to lead us to the new paradigm, let us all pick the same one. . . .
3. Once the move starts, the sooner we get it over with the better.

The herd instinct characterized by these three principles leads to a flashpoint change in the marketplace. That point signals the beginning of a tornado, which brings about rapid adoption of the new technology and sweeping change. The same kind of tornado hit New York after 1825.

Erie Canal Stories—Growth Along the Canal

The flourishing growth of this city must be attributed almost entirely to the Erie Canal, which has opened our internal Commerce with the West, and poured in upon us a rare, steady, and rapidly increasing stream of wealth.

Samuel B. Ruggles, 1832

As the canal was completed, it changed the experience of young people in New York. In particular, it encouraged a new and stimulating mobility. The entire population along the canal became familiar with new forms of river traffic—packet boats, line boats, freighters, and scows—and the commerce they brought. In every town people were employed to run the locks. Real estate near the canal took was developed to serve the traffic and people passing through.

In the early years, canal life reflected the sensibilities of the young and restless people who built and ran it. Farmers built low bridges over the canal to connect their lands. These were sometimes so low that passengers had to duck or be swept from the deck. According to reports and songs from the time, younger passengers found this exercise entertaining.

Oh, low bridge, ev'ry body down!
Low bridge, for we're coming to a town!
You'll always know your neighbor, you'll always know your pal
If you've ever navigated on the Erie Canal.

Not everyone was amused. An English visitor, Captain Basil Hall, soon tired of the fun. "It was rather amusing to hop down and then to hop up again," he wrote; "but, by and by, this skipping about became very tiresome, and marred the tranquillity of the day very much. There was no alternative but to return to the close and narrow cabin" (Walker and Walker 1963).

Canal commerce drove the development of western New York as goods and people traveled between New York City and the states and territories of the upper Midwest. The Erie Canal provided the cheap transportation that made it practical to ship wheat and other agricultural products from west to east instead of south to New Orleans.

Before the construction of the canal, New York City was the nation's fifth largest seaport, after Boston, Baltimore, Philadelphia, and New Orleans. The canal attracted Atlantic shipping to New York City. Within fifteen years of the canal's opening, New York was the busiest port in America, moving more freight than Boston, Baltimore, and New Orleans combined. It also became the main port of entry into the United States for immigrants from Europe. Many rode the canal boats to the rich agricultural lands of the Midwest, while others remained to swell Manhattan's population and fill the jobs created by the canal's economic impetus.

The canal also established an enduring population pattern for New York as a whole. The counties bordering the canal west of the Mohawk River were only lightly settled before 1820. Between 1810 and 1835 this region tripled in population, becoming more populous than the sections of eastern New York (except New York City) that were settled much earlier. Today, approximately 75 percent of New York's population still lives within the corridors created by the Erie Canal system.

Starting out as an object of dreams and a dreaded expense, the Erie Canal grew to become a vital part of the social fabric that shaped much of the growth of New York and the Midwest in the second quarter of the nineteenth century.

America's Information Utility

Welcome to the World of the Source . . . you are now part of the Information Age.
Users' Manual & Master Index to the Source, 1980

We asked ourselves. What were the main things that consumers would like to do if they had a computer in their home, other than word processing? . . . Put yourself back in 1978. There was no CNN. You didn't have 24 hour news. You had television, but television was 5:00 news, 6:00 news, and 11:00 news. You had none of this 24 hour stuff. So the idea of being able to access the UPI news wire as a consumer was a very exciting notion . . .
Bettie Steiger, 1998

In 1978 Bill von Meister had an idea. He had made money earlier by manufacturing telephone relays and was very aware of the rapid changes in communications and computing technology. His new idea was to create an information utility—information and information services offered online. What Vannevar Bush, J.C.R Licklider, and others dreamed about, von Meister decided to build.

He founded the Source in 1978. Initially called the Telecommunications Corporation of America, it pioneered many of the information services now available on the Internet. One of the most popular early features was the UPI news wire and its twenty-four-hour access to the news. The Source also offered electronic mail and pioneered the idea of on-line chat rooms, which were then called "seminar discussion topics." To send e-mail to another Source user, even one without a terminal, a user could telephone a special operator, who would key in the "voicegram" and send the message. An electronic catalog called DATA BUCKS allowed consumers to select from "an enormous array of products—from office machines to washers to television sets to living room furniture."

The core of the Source was its wide range of on-line information services. The 1980 index to the Source lists over six hundred subjects, and by 1981, the scope of available databases was even more comprehensive. Here are the top-level categories from the *Users' Manual* (1981).

INFORMATION SERVICES

- Source * Plus
- Business
- Catalog Shopping
- Education and Careers
- Government and Politics
- Home and Leisure
- News and Sports
- Science and Technology
- Travel, Dining, and Entertainment

Under these general categories many detailed services were offered. Under Business, for example, there were both reference sources and tools for managing a stock portfolio. Under Catalog Shopping users could chose from such specialized areas as classic radio programs, books, and classified ads. Under News and Sports they could find facilities for reading and searching the UPI news wire and *The New York Times*. Politically active users could use the Legi-Slate service to track House and Senate bills as they emerged from committee and moved through the process of debate, referral, vote, and signing.

The Source was very proactive in licensing databases. Initially unable to sign up the *Official Airline Guide (OAG),* the Source proceeded to sign up the guides of individual airlines, including, among others, American Airlines and United Airlines. Recognizing the implications of this trend, the *OAG* decided to sign up.

Early users of the Source came from all over the United States. Through their easy e-mail access to Source management they provided a constant stream of suggestions for new services and system features. Most were connected to the Source via terminals and modems through TYMNET and TELENET, which provided telephone connections to their computers. Data rates of the modems varied, from the established 300 bauds to the then-cutting edge 1200 bauds. As graphics were impractical at these speeds, information from the Source was delivered entirely in text form. Users were assigned ID numbers, initially tokens like "TCA123," reflecting the original name of the company.

The base of subscribers grew every year. By the mid 1980s, the Source had fifty-seven thousand subscribers. Reflecting on the history of the Source, however, we can see that it failed to develop a sustainable financial model. The start-up fee to join the Source was $25. After that, charges were based on communications costs and connect time. Expenses were higher than expected, for several reasons. One was management's decision to buy its own computers instead of renting night time on mainframes as originally planned. This change of course right at the beginning was motivated by the desire to assure reliable service through redundant and dedicated computers, but it was expensive. Another factor was that the Source's 1978 business projections were based on the unrealistic assumption that personal computers would be in every U.S. home in five years.

By the end of 1980, the Source was losing too much money. Von Meister was ousted, and the company was put up for sale. He went on to found GameLine, a company offering downloads of video games by Atari and others. Knight-Ridder, Cox Cable, and *Times-Mirror* all bid on the Source but needed stockholder approval to complete the deal. Before they could get it, *Reader's Digest,* a privately held corporation, stepped in and bought the company.

Reader's Digest management, worried that electronic publishing would have a negative impact on their existing revenue stream, nonetheless wanted to understand electronic publishing. The Source grew under its new management, but not as quickly as the competition. *Reader's Digest* was a relatively conservative company. When it first offered a book for sale, it did not print any copies until at least a hundred thousand orders were received. If fewer orders came in, the book was never printed. *Reader's Digest* managed the Source as a traditional publishing entity. Compared with other online services, it was not investing enough money into growing the business. The Source was overtaken by CompuServe and eventually sold to it.

From the perspective of the technology adoption cycle, the Source had crossed the chasm and developed niche markets. However, it never rode the tornado—partly because of bad management decisions and a lack of investment. But perhaps the key factor was that for users the expense of getting on line was still very high. Even though the Source developed special deals with terminal suppliers, modem manufacturers, and communications companies, users had to spend a significant amount of money to access the Source. For example, the Texas Instrument terminal that was most popular for accessing on-line services cost $1895. To join the Source, users had to buy expensive equipment they might not need for other purposes, and from which the Source itself realized very little income.

From the perspective of the Internet edge, the Source represents an early, bold attempt to "wire the world"—shrinking distance and distributing information to on-line communities. It pioneered the licensing of material for on-line distribution and showed many publishers that, contrary to their fears, on-line distribution actually promoted hard copy sales. The dream of using electronic information for education, a popular theme of Internet promotion in the late 1990s, was played out on a smaller scale on the Source. Jack Taub, an major early investor in the Source, convinced the Kellogg Foundation to let him wire the schools in Grand Rapids, Michigan, to provide on-line information access for every pupil. He saw the Source, as many today see the Internet, as a great tool for education.

Resisting Change

God created the universe in six days, and rested on the seventh—but He did not have an installed base to worry about.

Old joke revived by Giuliana Lavendel

Timing is everything. If you undertake these changes while your company is still healthy, while your ongoing business forms a protective bubble in which you can experiment with new ways of doing business, you can save much of your company's strength, your employees, and your strategic position.

Andrew Grove, *Only the Paranoid Survive*

After a technology rides the tornado, it becomes the dominant approach with a large installed base. The late majority continues to buy the technology and expand the installed base further. But nothing lasts forever.

Eventually, other, superior technologies come along and roles shift. Those who had promoted change by introducing and spreading a technology become the conservatives who resist the changes offered by the next technology

An installed base is both an advantage and a burden. It is an advantage because it can provide income to help a technology extend its time on Main Street. It can also be a disadvantage when the need for backward compatibility limits the technology's ability to evolve to meet new challenges.

Erie Canal Stories—Meeting the Railroads

[The railroads] . . . can reach no such point of supremacy, but where canals cannot be had they may be substituted, and prove, indeed, superiour to any other land conveyance.
Buffalo (N.Y.) *Journal*, 1831

From its opening in 1825 the Erie Canal dominated transportation in New York state, bringing with it a new prosperity and a new way of life. The first railroads built in New York were as used as feeder lines and seen as complementary to the canal. They would carry passengers on short feeder lines, and canal boats would carry the heavy freight. The first railroad construction began in the 1830s with the building of the Mohawk and Hudson line from Albany to Schenectady.

In the early 1840s, when railroads began carrying freight, the canal interests intervened politically. The state legislature required the railroads to pay canal fees and refused to issue them licenses to carry freight, except when canal navigation was suspended during the winter. Nonetheless, the railroads continued to grow. By 1842, the Attica and Buffalo Railroad, the Tonawanda Railroad, and six other lines completed the rails link between Albany in the east and Lake Erie in the west. By 1843, three daily trains ran each way across the state, charging an average passenger fare of three cents a mile.

Responding to this competition, the canal interests improved and enlarged the canal boats and widened the channels. The first boats had been towed by teams of horses from towpaths on both sides of the canals. In the 1840s, larger steam-powered canal boats were introduced. One builder, Seth Jones of Rochester, produced canal freighters ninety-seven feet long and seventeen feet wide with a carrying capacity of a thousand barrels of flour or four thousand bushels of wheat. The Canal Board lowered the tolls on freight. For a while this reduction, together with the greater efficiency realized with the larger boats, lowered the cost of transportation and kept the canal competitive with the railroads.

Throughout the 1840s the freight traffic on the canal grew. Over the decade freighter mileage increased from 5,556,950 to 11,733,250. The volume of freight carried was greater than that of all of the railroads combined. The Erie Canal seemed to be entering its golden age, yet the growth of the railroads foreshadowed its fate.

Ultimately, backward compatibility proved too great a burden. Canal freighters floated on water; reaching places away from the canal required new feeder channels, which were much more expensive to build than railroad tracks. Furthermore, the canal froze during the long New York winter, forcing the boats to stop for several months. By the 1850s, much of the country east of the Mississippi was crisscrossed with railroads. The faster service and greater connectivity of the railroads brought them dominance, and the Erie Canal gradually faded from prominence.

Upgrading Technology—Trying to Change a Rootball

> After the third try and the third set of guys went out the door failing to [re-architect] this thing it occurred to me that the only solution was to start over. It just couldn't be fixed. . . . It always reminded me of an organism. It was alive. It fought any attempt to change it.
>
> Jeff Crigler, 1997

Situations are always evolving. When a technology comes into general use, the market exerts pressures for greater efficiency and developers respond by making changes and optimizations intended to integrate it more deeply into the environment. Sometimes this integration becomes so complex that it threatens the ability of the technology to evolve any further.

Jeff Crigler is a technologist and leader in digital publishing who has been to the edge several times while developing computer systems for information delivery. A company he helped found—Lobbyist Systems Corporation—was purchased by Lexis-Nexis. Lobbyist Systems specialized in providing its clients up-to-date information about members of

Congress—their voting records, major contributors, and personal data. The information was collected at a central point and sent to client sites on a daily basis by telephone line. At the time, with personal computers still in a nascent stage, mainframe computers were the dominant technology. Enjoying all the usual advantages of an established technology, mainframe interests like Lexis-Nexis were able to marshal their greater marketing force to take over the market for electronic document and data distribution.

In 1987 Lobbyist Systems was purchased by Lexis-Nexis and Crigler became a director in product development. Lexis-Nexis is a comprehensive system of research databases offering on-line information in journalism, law, and business; it includes full texts of mainstream publications, industry newsletters, directories, and research reports. The buyout gave Crigler the opportunity to observe the mainframe computer technology overcome an approach based on personal computers and, later, to watch the tensions that arose when engineers tried to upgrade a system that had a large installed base.

Lexis-Nexis was founded in the late 1960s. At the time Crigler left the firm in 1995 its thirty-two s/390 mainframe computers with six terabytes of mirrored data—all on fast direct-access storage devices that could handle two thousand simultaneous queries—had become an enormously complex system.

> Everything had been rewritten. There was nothing in it that had not been tuned to support the environment. And it was absolutely impossible to change. Every possible change threatened to cascade. You front-end the front-end to add that next level of capability. You don't dare touch anything underneath because of consequences that you can't see at all. When you wind up layering the onion like this, what happens is that the system does not change. (Crigler 1997)

In the parlance of computer design, the Lexis-Nexis system of the mid–1990s had grown in a way that violated *modularity*—the idea that individual components have relatively narrow and constrained interfaces so that they can be replaced or upgraded independently. Without modularity, a system becomes a dense rootball of interconnected parts. It is difficult to change because it is difficult to understand the consequences of making a change. Crigler saw hundreds of millions of dollars being spent, unsuccessfully, in attempts to re-engineer and re-architect the system.

In 1995, Crigler left Lexis-Nexis to join IBM and to lead and create the IBM infoMarket™. As Crigler puts it, he had an epiphany about where the information industry was heading. He believed that big, centralized on-line services would have a difficult time competing on the Web. The infoMarket was created to act as an intermediary for publishers without interfering in the relationship between publishers and their customers. It provided services (e.g., delivery services) and collected for those services. Part of the technology push of infoMarket was because of IBM's Cryptolopes™, a kind of digital envelope for delivering content (see chapter 3), and its successor technology, CryptolopeLive™.

CryptolopeLive™ was designed to manage gracefully the problem of *legacy systems* for a distributed system. Legacy systems are installed systems you need to connect with, and which are difficult to replace or change. In secure digital distribution, a digital work is encrypted, placed in a digital envelope, and sent to a user's computer. In many architectures, the user's computer must contact the publisher's computer from time to time to record use and receive billing and updating information. With CryptolopeLive™, such communications are facilitated by software shipped in the envelope with the digital work. But what happens when the publisher's Net server changes? In an "onion-layer" approach, the publisher would need to keep on line all systems that are compatible with all the digital works it has ever shipped. With CryptolopeLive™, the software shipped with the digital work "notices" that it is outdated and contacts the on-line server to replace its outdated modules with new modules from the publisher's site. Thus, instead of building more layers on the onion—building front-ends to front-ends at the server—this approach essentially refreshes the outer layer of the onion.

Reflections

Our computers have reached the point where to do all the applications any reasonable person wants; it wouldn't cause any threat if the technology were frozen for ten years. The way computers are advancing now has had the wonderful consequence that we keep getting very inexpensive platforms with ever widening capability, and I wouldn't want to stop that. . . . But there's just a lot of people in the world who can't afford it. We ought to come out with the computer equivalent of a Ford Model T or a Volkswagen. This would reduce greatly the maintenance costs, not only in terms of the replacement hardware and software, but also all of the configuration you have to do.

Joshua Lederberg, 1997

During the early stages of the technology adoption cycle when a technology is emerging, its promoters struggle for change against strong resistance. In this early market stage the promoters are dreamers, innovators, and early adopters. Initially, resistance stems mainly from technological shortcomings and the challenges of integration. At the chasm stage, if the challenge to go beyond the early adopters to a broader market is met, the bowling pin and tornado stages follow. At that point, the advantages of a technology for growing numbers of people create a force for change and resistance wanes as older technologies are displaced. If successful, a technology eventually becomes the new dominant technology with a large installed base.

This installed base is both an asset and a liability. It is a liability because it has inertia that favors backward compatibility but inhibits change. But nothing lasts forever. Depending on the circumstances, a technology's advocates may be able to leverage their superior market position to change with the times. Or they may become locked into decreasingly workable approaches.

When we understand the dynamics of the technology adoption cycle, predictions about the rise and fall of technologies seem inevitable. Evolution follows revolution. Once a technology is widely deployed, it is expensive to throw it away too soon. During this time, the cycle favors backward compatibility and evolution of the technology. Ultimately, though, the limits of its ability to evolve are reached and it is replaced. The next revolution happens when the next technology arises.

Global Challenges of Rapid Change

The dynamics of rapid, incremental change are epitomized by progress in integrated circuit technology, which for over thirty years has been characterized by Moore's Law. This rapid rate of change in computing devices has left consumers with two fundamental choices: what level of performance to pay for and how often to replace their equipment. In terms of level, they can choose to buy the hottest equipment at the leading edge of technology introduction, ride the middle wave, or go with the discounted tail. Prices of systems from one level to the next increase by a factor of two to three.

Within each level, the price of a computer stays relatively constant but hardware performance improves.

It is not uncommon for families to keep older computers around as long as they are still working. At the time of this writing, the oldest computer in my home is an 8-MHz Macintosh™ LC made around 1987. The years since then, with Moore's Law compounding at 44 percent per year, have yielded state-of-the art computers that make that old computer seem unimaginably slow and outdated. Yet, although too slow for browsing the Net, with software of the same vintage it is still good enough for word processing and exchanging e-mail.

This observation about the usability of old computers and Lederberg's call for a Model T version of the computer raise questions about the value of the seemingly frantic pace of change in computer technology. On the other hand, we can view rapid technological change as the necessary engine of creation and innovation. By buying state-of-the-art systems, a small fraction of the world's computer users are paying for the development of arguably more capable systems every year.

The argument that the software running on these systems is delivering a corresponding increase in useful power is much more difficult to sustain. According to industry lore, hardware engineers lament that every advance in the power of hardware is eaten up by a corresponding "advance" in software. Cynical observers, however, have noted that bloated software and artificial incompatibilities between software releases actually help drive the market for new hardware by effectively making older systems obsolete.

The cost differential between computers at the state of the art and computers that are five years old creates choices that affect the accessibility of computing for most of the world, which cannot afford the latest systems. Corporations in wealthy advanced nations can afford a more rapid pace of change than those in the poorer, developing countries.

What are the implications of this inequality for the Net? Will the Net become more fragmented, increasingly separating the information haves from the information have-nots? Alternatively, will concerns about backward compatibility for a growing installed base of old systems across the world create inertia that stifles innovation? We consider these questions and the issues surrounding them further in chapter 9.

Waves of Change and the Digital Cambrian

Although she takes many different forms, this goddess—sometimes a Black Madonna or an Asian or Indian Madonna—always carries authority. . . . Living in the creative intercourse between chaos and order, she calls us to enter into the dance of creation.

Marion Woodman and Elinor Pickson, *Dancing in the Flames*

The "Digital Cambrian Age" described in chapter 2 puts an interesting twist on these questions. The Net is not just one technology or one thing. Rather, it is made up of many technologies, and it is evolving to serve several functions at once. Waves of different competing and complementary technologies are moving forward simultaneously but at different stages, getting stuck at chasms, and sometimes reaching Main Street.

The technology adoption cycle focuses attention on one suite of technologies at a time. Looking back at the Erie Canal through this lens, we saw it first rising and then falling in prominence. However, viewing the era through a wide-angle lens, we can see waves of transportation technologies rising, falling, and co-evolving—dirt roads and trails, rivers, the Erie Canal, the railroads, and the highway system. Just as the canal was overshadowed by the railroads, the railroads were later overshadowed by trucks on the interstate highway system. Displacement in both cases was driven largely by the advantages of greater connectivity. Today, these technologies constitute New York state's transportation infrastructure What is left of the Erie Canal is a historical artifact used mostly for recreation.

The fate of every technology depends on its environment, which is different in different parts of the world. The rates of adoption of the cellular telephone in various parts of the world provide a well-known and illuminating example. Where there was a large installed base of wired telephones (as in the United States), adoption of cellular phones took many years. But in places with a pent-up demand for telephones and slow expansion of the wired service, cellular phones rode the tornado to Main Street rapidly and vigorously.

By analogy, we should expect different technologies at the Internet edge to emerge more rapidly in some parts of the world than in others. Perhaps, like cellular telephones, PDRs with broadband communication will ride a rapid tornado in countries and regions lacking an installed base of wired

Net connections. The rapid evolution of portable networked devices suggests a global future that includes waves of devices for digital services quite unlike the deskbound systems that dominated the Net in the late 1990s.

The Internet edge has a complex dynamic. To paraphrase the quotation from Arnold Mindell in chapter 1, an edge marks the limits of who we are and what we imagine ourselves to be capable of. It represents an identity crisis. The Net accelerates change as it connects us together. The tribes of our planet are now dancing at the edge of chaos, inventing and expanding the Net. It will surely serve many different interests and become a force for both convergence and diversity. What the Net becomes, and what we become, is deeply involved in a process that challenges our abilities to understand each other.

8

The Digital Keyhole: Privacy Rights and Trusted Systems

For now, consider everything you do on the Internet completely public. Your illusion of privacy surfing the net from your own bedroom late at night with the door closed is just an illusion. The FBI might not be there tomorrow, but your risk of exposure lasts for years and years.

Dick Mills, 1996

It is not easy to observe closely, to take the time and to make the subtle moves that allow the soul to reveal itself further.

Thomas Moore, *Care of the Soul*

Privacy itself is nothing new. The dream of finding quiet and private spaces for some of life's activities probably goes back beyond the historical record. In England the legal precedent that "a man's home is his castle" dates from at least 1604; in Semayne's Case the judge ruled against the officers of James I who had barged into a house without announcing themselves.

During the 1970s, people in many countries around the world recognized that computer technology was shifting the balance regarding privacy. It was becoming clear that computers and computer networks are efficient instruments for gathering and correlating information. They had become progressively more common since the 1960s and were being used by governments, private enterprise, and individuals to keep track of people. The pervasiveness of computers and networking have challenged our societies to come to terms with a question that seems both important and elusive: With technology evolving faster than our collective understanding of its effects, what aspects of privacy are threatened and is it is possible to protect them?

In many ways the privacy issue is analogous to the copyright question discussed in chapter 3. Just as copyright law is concerned with the use of published works, privacy law is concerned with the use of private and personal data. In both cases, technological developments are overturning established protections, expectations, and practices. Programmed in conventional ways, computers can radically undermine copyright and privacy by allowing the unauthorized use of copyrighted works and private data.

In both cases, the ubiquity of computers and the need for checks and balances in an evolving situation has made crafting an appropriate response—whether legal or technological—difficult and often puzzling. Making laws is one thing, but having effective ways to enforce them in the digital world is quite another. There are too many gray areas, and people sometimes find it onerous to comply exactly with the law. Moreover, copyright and privacy laws are themselves evolving; users, copyright holders, equipment manufacturers, service providers, policymakers, and others are discussing new rules, policies, guidelines, and standards. There are competing social values at play, and competing interests that will determine just what such laws will say. At the same time, opportunities for using technology to enhance both privacy and copyright by designing computers as trusted systems are emerging. To achieve privacy at the Internet edge we will need to combine technical sophistication with wisdom.

What Is Privacy?

Information privacy is not an unlimited or absolute right. Individuals cannot suppress public records, nor control information about themselves that, by law, is used for a permissible purpose (e.g., criminal defendants cannot prevent courts from examining their prior criminal record before imposing sentence, and sellers of realty cannot prevent a title search of their property). . . . As a practical matter, individuals cannot participate fully in society without revealing vast amounts of personal data.

National Information Infrastructure Task Force, *Options for Promoting Privacy*

A little paranoia . . . isn't all that crazy. . . . If you have lots of sensitive material, don't saddle yourself with a lightweight cut-up. Value America has GBC models that will shred up to 19 pages at a time, at a blazing 45 feet per minute . . . even enough capacity for the White House.

Advertisement for paper shredders, *Wall Street Journal*

What is privacy? Who needs it? These questions are not just rhetorical, for privacy legislation and debate have recently had to consider several different notions of privacy. The questions are germane to the Internet edge because computers and networks are changing the balance and can either diminish or enhance privacy. One way to get some perspective on what personal privacy means is to look at several examples of it.

Many people value time spent alone. With no one watching, no one to please, no one to interrupt our thoughts, and no one to respond to, we can relax. Without distractions we can notice life's rhythms. Such time alone is what some teachers and mystics call soul time. Soul time builds equanimity and energy. Although it is possible to have soul time without privacy—while the phone is ringing, people are knocking at the door, and strangers are watching us in public places—it is more difficult. Private time can also include time spent secluded with friends and family. Alone with close friends, we feel differently, act differently, and are less on guard than we are in public.

Privacy issues also include protection for records—private communications, whether written on paper, faxed over a telephone line, sent by e-mail, or recorded on an answering machine. Convention, and to some extent the law, assure us that when we mail a letter to a loved one, it will not be opened by a third party. A paper envelope provides a barrier to clandestine readers so that skill and effort are required to steam it open and then reseal it. However, the real security of a paper envelope derives not from its flimsy physical barrier but from the rules of a society that values privacy. Similarly, our personal records—whether sentimental, medical, or financial—may contain sensitive information that we do not want to be available for public viewing and comment.

We value our privacy, among other reasons, because we believe strangers who know too much about us could exert improper control over our lives. They could take information out of context and use it against us. Thieves who know our vacation plans could time their visits to our homes; gossips who know some of our personal business may act to tarnish our reputations; advertisers who know our interests could deluge us with unwanted sales promotions. In short, personal privacy is about freedom from surveillance, freedom from intrusion, and security for our personal records. All these themes of privacy concern preventing other people from monitoring our activities in detail.

Privacy Cues and Cyberspace

Computers and networks are shifting the grounds of privacy. In the real world, physical cues such as closed doors, locked cabinets, and sealed envelopes remind us to respect privacy. They mark the boundaries between public and private spaces. In cyberspace, the physical cues are lost, and distinctions between public and private space are blurred.

The conventions and expectations of privacy we have in the real world do not always carry over to cyberspace. For example, we feel sure that a letter sent in the mail will be opened only by the recipient. We may expect that to be true in cyberspace too, but it isn't necessarily. Federal law does not apply to e-mail, and unencrypted e-mail can often be read by third parties. Indeed, some corporations have asserted a legal right to read their employees' electronic mail.

When we look at people in the real world, they can generally see us too; vision and line-of-sight work both ways. By contrast, the technology by which cameras are networked in cyberspace can let other people watch us without our knowledge—a phenomenon many people find very disturbing.

Computers and Data Protection

It should come as no surprise that advances in computers have led to issues about privacy. Computers are designed to process information efficiently, and networks are designed to transport it efficiently. Together they are ideal media for gathering information from many places, correlating data and stories from different sources, and assembling the related pieces into a dossier on a person or company.

Legislating Data Protection

But where does one "personal information system" end and another begin in the more decentralized and networked "information highway" environment?
Colin J. Bennett, "Convergence Revisited"

Recognizing that computers and computer networks are changing the balance regarding privacy, many countries have developed new laws. The literature on privacy legislation and the different approaches adopted by national governments is broad. We can, at best, only survey some the issues it raises by focusing on data protection. Readers interested in more information will find several sources in the Further Readings section at the end of this book.

In the United States, there are three main sources of privacy law: constitutional protections, common law or case law (precedent), and statutory law. Although all three contribute to the legal basis of privacy, most of the action related to electronic information has been in statutory law. Most other countries have also chosen to enact new laws, especially in the area of data protection; worldwide the rapid development of computer technology has raised a series of challenges that have reshaped the legal contexts and prompted still-evolving legislative responses.

For the most part, the Constitution and subsequent Supreme Court decisions govern the powers and limitations of government, while the activities of private citizens and organizations are regulated by statutory law. Although the word *privacy* does not appear in the U.S. Constitution, several of its provisions bear on it. The Fifth Amendment, for example, states that no defendant in a criminal case shall be compelled to be a witness against himself or herself. Under this amendment, a person who wants to keep hidden information that is stored "in the mind" cannot always be compelled to produce it. This provision does not, however, cover personal information in written form, like letters or a diaries. Instead, such personal effects are protected by the Fourth Amendment, which states that "the right of the people to be secure in their persons, houses, papers, and effects, against unreasonable searches and seizures, shall not be violated." Various cases tried before the Supreme Court, such as *Katz* v. *United States* in 1967, have broadened the interpretation to include protection against wiretapping. The court has ruled that, even without physical trespass, wiretapping constitutes an illegal search under the Fourth Amendment. Thus, the amendment has been interpreted in a way that protects people and not simply places. Over the years, new laws and judicial decisions have refined the criteria and procedures defining when wiretapping is legal.

The Threat of Centralized Databanks

Without computers, a modern welfare state could not operate. This explains the thinly veiled euphoria of the bureaucracy for the new technology.

Viktor Mayer-Schönberger, "Generational Development of Data Protection in Europe"

Privacy legislation has evolved since the 1970s, when computers that seemed to fulfill the dire predictions of George Orwell's *1984* first appeared.

As plans for using large computers in government bureaus took shape, many people became concerned that the data banks they contained would be used by powerful and inhumane bureaucracies. The amalgamation of information would, they feared, create more centralized power and be used to advance the narrow goals of bureaucrats rather than those of society in general. Legislators, assuming that information would be kept in functionally specific, bureaucratically controlled databases, responded to the threat by creating regulatory institutions. Germany and Sweden, for example, established commissions and boards empowered to investigate compliance with data-protection norms.

In the United States, the first federal legislation specifically regulating the use of personal information was the Privacy Act of 1974. Its first principle is that the federal government should have no secret systems of record keeping. The act requires each agency to publish a notice of all the record systems it maintains and to set up procedures enabling U.S. citizens—but not foreigners or even resident aliens—to gain access to their records and make corrections. There were reportedly about five thousand government data banks in the United States at the end of the 1990s. Even a casual search of the Net for "Privacy Act of 1974" turns up hundreds of postings.

The Privacy Act also mandates that only information relevant to a specific purpose can be collected and that it be accurate, complete, and up to date. Another principle forbids external disclosure of an individual's personal data without the consent of the subject or a judicial authority. However, as the act includes no specific enforcement or oversight provisions and lacks a statutory test for judging disclosures, it leaves disclosure policy up to the agencies themselves. An agency can decide that a disclosure is compatible with the purpose for which a record was collected if it establishes that such disclosures are a "routine use." Privacy advocates have argued that agencies' ability to make such declarations of routine use without external oversight is a major loophole in the Privacy Act.

The Flow of Personal Data Across National Borders

By 1980, governments recognized that national differences in privacy legislation are economically important, because of their potential to interfere with the flow of information and, thereby, with international trade. The

Organisation for Economic Co-operation and Development (OECD)—successor to the Organisation for European Economic Co-operation (OEEC), which coordinated the rebuilding of Europe under the Marshall Plan after World War II—had earlier broadened its membership to include the United States, Canada, Japan, and six other countries. The OECD's 1980 guidelines for privacy and the flow of data were an attempt to harmonize members' privacy laws and advise countries that had not yet enacted such legislation. The guidelines became the basis of many national privacy laws as well as many voluntary codes of conduct.

To remove or avoid creating unnecessary obstacles to the flow of personal data or economic information, the OECD guidelines proposed the following principles, abbreviated here, to define members' minimum consensus on privacy policy.

- *Collection Limitation.* There should be limits to the collection of personal data; any such data should be obtained by lawful and fair means and, where appropriate, with the knowledge or consent of the data subject.
- *Data Quality.* Personal data should be relevant to the purposes for which they are collected and, to the extent necessary for those purposes, should be accurate, complete, and up to date.
- *Purpose Specification.* The purposes for which personal data are collected should be specified not later than the time of collection; subsequent use of the data should be limited to the fulfillment of those purposes or of such others as are compatible with them and as are specified each time the purpose changes.
- *Use Limitation.* Personal data should not be disclosed, made available, or otherwise used for purposes other than those specified, except with the consent of the subject or by the authority of law.
- *Security Safeguards.* Personal data should be protected by reasonable security safeguards against such risks as loss or unauthorized access, destruction, use, modification, or public disclosure.
- *Openness.* There should be a general policy of openness about developments, practices, and policies with respect to personal data. Means of establishing the existence and nature of personal data held and the purposes of their use should be readily available, as well as the identity and usual residence of the data controller.
- *Individual Participation.* Individuals should have the right to obtain from a data controller, or other official source, information or confirmation of whether data relating to them has been collected; they should receive, on request, copies of such data within a reasonable time frame at

a charge, if any, that is not excessive and in a readily intelligible form. They should be able to challenge any of the data and, if the challenge is successful, have the data erased, rectified, completed, or amended.
• *Accountability.* A data controller should be accountable for complying with measures that give effect to the principles stated above.

To a large extent these principles were grounded in the experience of the data-protection laws of the 1970s. What the OECD brought to the privacy debate was the realization that national privacy policies had an international dimension. Many countries, including the United States, endorsed these privacy principles.

When the OECD was crafting its guidelines, it recognized that balancing opposing interests in privacy legislation was unlikely to be accomplished quickly or once and for all. It acknowledged several areas in which members had difficulty reaching agreement, including: the nature and requirements for control and enforcement mechanisms, the application of privacy laws to corporations or "legal persons" as well as physical persons, and differences in attitude about which categories of data were especially sensitive. The overall mission of the OECD was to promote privacy legislation that would minimize disruption of trade and to provide a forum for ongoing debate of the privacy issue.

Informational Self-determination

By the 1980s the continuing development of technology had undermined the assumption that regulating data banks would control data use. Computers were getting smaller, faster, and more ubiquitous. Departmental computers, minicomputers, and even personal computers had become more common. Centralization was no longer seen as the main technological challenge, since data could be brought together on small computers and correlated at will. The second generation of European privacy laws focused instead on citizen rights. The constitutions of Austria, Spain, and Portugal were amended to provide citizens with informational privacy rights. New statutes were passed in France, Austria, Denmark, and Norway. Legislators assumed that these laws enabled citizens, supported by legally established privacy rights, to ensure their own privacy. Individuals could not only access and correct their personal data, they could also have a say in how data were used. For example, the Norwegian Data Protection Act gave individuals the right to forbid use of their data for direct marketing and market research. Privacy institutions were evolving from regulatory boards to bodies that could rule on disputes over a particular individual's rights. The French Privacy Commission, for example, is responsible for setting up registration procedures for data processing; it can also rule on disputes in individual cases. In Germany a 1983 case before the Constitutional Court popularized the phrase *informational self-determination,* referring to the individual's absolute right to decide on the release and use of his or her own personal data. Thus, in several European countries—including Germany, Austria, Norway, the Netherlands, and Finland—the philosophy behind legislative changes was to bring the individual into the loop for making decisions about personal information.

During 1980s European legislatures with were much more proactive than the U.S. Congress in relation to privacy legislation, on two counts. The first is that European laws were designed to empower the individual to seek specific privacy rights over particular kinds of data, no matter where the data records were kept. In other words, laws in Europe sought to protect the rights of citizens rather than trying to regulate specific data banks. Secondly, the new European laws governed private as well as government records. Legislatures in the United States have been reluctant to create an omnibus law to apply to private-sector as well as governmental data banks.

Unalienable Privacy Rights

By the 1990s experience with these laws showed that, even in Europe, the power was stacked against the individual. Few individuals could afford the financial and time commitment required to determine how information about them was used or to demand redress. Under the principle of informational self-determination, the forfeiture of privacy rights—such as might be specified in the small print of telephone or credit card contracts—was legally valid. Unfortunately, experience has shown that individuals purchasing information services easily give up privacy rights without even realizing that they had done so or that a choice was available.

Moving to redress the power imbalance between the citizen or consumer and large organizations, more recent European laws have sought to define certain privacy rights as unalienable and to provide individuals in disputes with professional help (such as from ombudsmen). Legislation in several European countries has specified, for example, that rights to control the release of credit records cannot be bargained away. Roughly speaking, recent European laws about such sensitive records have established unalienable rights for individuals and enacted norms for data processing.

Although the U.S. experience during this period has some similar elements, the common law used to decide many privacy issues has tended to lack teeth. An important criterion in judging proper use of data in a court case is the "reasonable person" test. According to this standard, because computers are now so widely used to collect and combine data, a reasonable person would have a rather low level of expectation about privacy. The privacy battle is thus lost before it starts. Case law is not likely to induce a record keeper to follow provisions like those of the Privacy Act—that is, to publish descriptions of records, limit data collection, meet quality standards, allow individual access and correction, and restrict internal uses of data.

During the 1990s, lacking effective common law and constitutional protections and unable to create a consensus for omnibus statutory regulation of data banks, the U.S. Congress enacted federal privacy laws to regulate use of several specific kinds of private data, including video-rental records, telephone records, cable-viewing records, and educational records. The Fair Credit Reporting Act of 1970, revised in 1996, mandates consumer access and correction as well as limits on data use and disclosure. Unlike earlier legislation, it spells out legal remedies and provides for administrative enforcement. In the late 1990s, however, use of other sensitive records, such as medical records, remained essentially unregulated.

The piecemeal U.S. approach parallels a recent shift in European law away from the omnibus regulation of data banks and informational self-determination and toward guarantees of specific unalienable rights over specific kinds of data.

European Regulatory Harmonization

For the past two decades an important activity of the European Commission has been the effort to harmonize the data-protection laws of the European Community (now European Union, EU). Because it has direct consequences for individuals and corporations from other countries doing business in Europe, the influence of this harmonization process has extended worldwide and has prompted public debate and awareness of privacy issues.

In 1981, the Council of Europe issued a convention for the protection of individuals from misuse of personal data resulting from automatic processing systems. Among other provisions, this convention lays out principles of data protection and directives for transporting personal data across national borders. The convention is somewhat abstract; it lays out general principles for defining categories of data, data security, safeguards, sanctions, and such things but leaves specific details of implementation and enforcement to EU member states. Moreover, in principle, member states can, after giving notice of their intention, decide individually not to follow the convention. The convention, which affects any corporation doing business in Europe, went into effect in 1985 after it was ratified by France, Germany, Norway, Spain, and Sweden.

In 1995, the European Commission issued the European Directive on the Protection of Personal Data—95/46 EC—which took effect in October 1998. The directive has several significant effects on record keeping. For example, each member state must establish specific authorities responsible for monitoring data privacy in its territory. When an organization collects data from a subject, it must provide to the authority the identity of the controller, the purposes of the information processing, and information about data recipients. In addition, it must indicate which of the data obtained are voluntary and which are obligatory. Finally, the directive requires the data controller to disclose whether the subject has the right to review and correct his or her personal data. Combined with the 1981 convention, the directive acts to block transfer or exchange of personal data not collected in accordance with EU standards.

In many European countries—and to some extent, in the United States—privacy law has evolved toward the view that individuals possess certain unalienable rights to privacy from the unchecked collection, storage, and distribution of personal data. Under the Freedom of Information Act (1982), U.S. citizens may request access to any written or electronic records held by the federal government. Furthermore, the Privacy Act of 1974 regulates many aspects of the federal government's record keeping. Although there is no omnibus regulation covering records held by private parties in the United States, certain classes of data, such as credit records, are regulated. Privacy advocates in the late 1990s often note, however, that the United States, unlike member states of the European Union, has no oversight agency for privacy enforcement.

Commerce and Intrusion

We can look at your customers and tell you a lot more about them. More than you ever thought possible. And not as a group, but as individuals. By exact age. Sex. Income. Lifestyle characteristics. Life event. And more. And we can help you decide the most effective ways to use this type of information to achieve your marketing goals.
Advertisement for Metromail, Inc.

In addition to legislative actions, there are several policies or policy frameworks for regulating private data developed by organizations in the commercial sector of the United States—for example the Information Technology Industry Council (ITIC) and TRUSTe. Rather than mandating a particular standard, both organizations recognize that personal data varies in levels of sensitivity and in the need for protection. They thus advocate a self-regulating approach in which each industry group works out a general policy framework that member organizations can adopt; each member or licensee company then develops a policy whose specific elements are consistent with the general framework. Establishing a broad policy framework and letting participating companies tailor their own particular policies is easier than obtaining agreement on specific elements. However, lack of a uniform code leaves to the concerned consumer the work and responsibility for checking the details of an individual company's policy.

In addition, neither ITIC's nor TRUSTe's policy framework covers the issue of sharing collected data with other parties. TRUSTe's framework does, however, differ from the ITIC's in requiring licensees to cooperate with TRUSTe reviews and audits of their policies. However, as most privacy policies from commercial entities lack independent audit and sanctions provisions, it is doubtful that such policies can exert enough influence to head off legislative action.

Another organization, the Information Technology Policy Council (ITPC), is a group of major high-technology trade associations that includes, among others, the American Electronics Association, the Business Software Alliance, the Computer and Communications Industry Association, the Computer Systems Policy Project, the Consumer Electronic Manufacturers Association, and the Telecommunications Industry Association. In the spring 1998 it proposed an eighteen-month plan for adopting and implementing an industry-led, self-regulatory scheme to address privacy concerns in on-line applications. The "ITPC Industry Self-Regulatory Privacy Plan" expresses its members' agreement on the need for a unified set of privacy principles as a model for the high-tech industry. Further, it states that these principles should be codified in such a way as to identify alternative avenues for consumer recourse if a company adopting them violates the guidelines.

In the late 1990s awareness of issues relating to data privacy in the United States expanded to include many different kinds of privately held records. The power of computers and networks has created risks to privacy because so many ordinary things we do in the course of a day leave electronic trails.

When we use a telephone, a record is made of whom we called or who called us. If we carry a cellular telephone, the cellular network automatically keeps track of where we are so as to create a connection if someone calls us. Whenever we use a credit card or an ATM card, a record of the place and time of the purchase and, often, the goods purchased, is made. If we use a supermarket discount card, a record is kept of exactly what items we have purchased. Our credit histories, including our current credit card balances and any credit disputes we are involved in, are maintained on-line and are managed and sold by credit bureaus. The structure of the credit-reporting business is likely to err on the side of overstating potential credit problems, even when the information in the data base is outdated or incorrect.

When we surf the computer network, some programs keep track of the places we visit. In principle, it is possible to identify every file we read and how long we tarry over a news story. Most people are not consciously aware of these capabilities. If, for example, a woman reads a digital magazine on fishing, should computer systems across the country record this interest and deluge her with special offers for fishing gear and fishing magazines? If a man with a heart condition develops an interest in deep sea fishing, should his insurance rates go up because of the risk he will cost the company more money if he has a heart attack at sea?

A particular concern many people have about computers and networks is the way they simplify the process of merging data from different sources. For example, in 1996 someone in Canada proposed to catch unemployment cheaters by matching the customs forms of Canadians returning from a journey to vouchers for unemployment insurance—a purpose having nothing to do with identifying the value of goods brought from abroad. This is just one example of the many ways that a piece of information, a name or a personal identity number (like a social security number), can make it easy to assemble data from diverse sources. When data are routinely collected for one purpose they can be routinely used for another. According to some privacy advocates, this practice violates the principle that data should be collected and used only for a specific purpose.

In 1988, to address concerns that some federal agencies were employing the routine-use exception to justify widespread electronic comparison of federal databases, the U.S. Congress passed the Computer Matching and Privacy Protection Act. The new law, which amends the Privacy Act, regulates the agencies' use and exchange of the information in their data bases and specifies procedures for making automated comparisons. Individuals who suffer adverse consequences as a result of such data exchanges—such as loss of a government benefit—must be notified and given an opportunity to refute the adverse information. The act also requires each agency to appoint a board to oversee its use of data comparisons.

As a practical matter, in the late 1990s our most sensitive information was not available on-line without restrictions, at least in the United States. However, in the private sector, paying a modest fee substantially increases the amount and quality of potentially sensitive information about us someone can obtain. Thus, even though freely available mundane information such as telephone numbers is often incomplete and out of date, anyone with $6.95 can search on-line for bankruptcy records and information about lawsuits. For about $20 other services on the Net will provide someone's current address and telephone number, a list of neighbors, and even a social security number. Information is also available about people's personal assets, such as real estate, stock, aircraft, and boats. This is the so-called second tier of information. In the United States, information in the third tier, which includes credit data and earning history, is restricted; the Fair Credit Reporting Act limits access to individuals and organizations that can demonstrate a "reasonable" or official need for the information.

In the private sphere, it is quite easy to merge data from various commercial data banks. People who order goods by mail often notice an increase in solicitations from other mail order houses. People who accumulate a significant credit debt are likely to receive numerous offers from rival credit card companies and home finance companies eager to encourage more borrowing, spending, and interest payments. European laws explicitly point to direct marketing as one of the reasons why people might object to governmental release of their personal data.

As our speculation about sponsored browsers in chapter 2 suggests, the future may hold even more extreme uses of merged data. As we increase the amount of information and entertainment we receive through digital devices, it becomes more and more possible to collect detailed data about our reading and viewing habits. In principle, it would be quite easy to develop systems and programs that notice what we read or watch, when we read or watch it, and how long we spend doing so. A profile of our interests could be generated from such records, making it easy for advertisers to target us with junk mail, junk e-mail, or even personalized television advertisements. Net browsers already target our interests in this way. It can be amusing to perform a Net search, first for gaming or gambling, and then for (say) IBM laptops—and watch how the banner advertisements at the bottom of the screen change to track the apparent shifts in our interests. Depending on many factors, these banner ads can be perceived as conveniences, annoyances, or evidence that our browsing activities are being monitored.

It is clear that our everyday activities leave a trail of information about what we are doing and that this information resides in cyberspace. Advertisers—and anyone else who feels like it—can use computers and networks to pull together those little bits of data into a picture that may be much more revealing than we realize or appreciate.

Surveillance and Privacy

Why is it acceptable for passersby to stare in through the window of a restaurant at the people dining there, but not through the window of a house to see what the occupants are up to? Why do people try on clothes in a fitting room, but not in other parts of a store? . . . In public and private places there are different more or less implicit rules about acceptable behaviors and interpersonal access rights. . . . People learn these rules in the course of normal socialization.

Victoria Bellotti, "Design for Privacy in Multimedia Computing and Communications Environments"

Security in *1984* isn't that much different from security now. Just about every grocery store, business, parking lot, and airport has security cameras watching you. . . . Do we need such a lack of privacy just to catch some criminals?

Morgan Stefik, 1997

Modern life is increasingly being recorded. In San Francisco and many other cities, automatic cameras at certain traffic intersections photograph automobiles driving through red lights. Parking garages, banks, and retail stores routinely employ video cameras to provide evidence of crimes that may be carried out in them. Most people who live in urban settings are recorded on video several times a day, while walking through banks, stores, train stations, shopping malls, and parking lots. As noted in chapter 2, many of these cameras are integrated into computers and networks.

Nor are all the video cameras owned by government or commercial organizations. In the late 1990s, approximately nine million video camcorders were sold each year. The prevalence of these devices and the activities of aggressive photographers with telephoto lenses have led to a public perception that personal privacy is being undermined to create bogus news stories.

Since the late 1980s, researchers interested in increasing the effectiveness of collaborative projects have experimented with remote video, audio, and location-sensing technologies. This research, which has also raised issues about privacy in the workplace, has been conducted in both commercial firms (e.g., Apple Computers, Bellcore, Sun Microsystems, and Xerox) and academic settings (e.g., University of Toronto, Massachusetts Institute of Technology, and the University of Michigan). One application of computer-based surveillance is *activity-based retrieval,* which uses the record of a person's activity to index and retrieve documents. For example, one could ask such a system to "retrieve the document Mark e-mailed me after he visited last month." Video technology has also been used to create collaborative workspaces that support conversation and information sharing among people in distributed work groups. It has even been used to create a new kind of social browsing by mounting a video camera at the company coffee machine; people at their work stations can see who is there on their computer screens and go to meet them for conversation.

Privacy-aware designers of these systems have noticed that there is great variability in people's views of when privacy has been invaded. At Xerox PARC, several systems provide video access to others in the Center and to Xerox laboratories in other parts of the world. Some people working on a project together maintain an almost constant video connection between their offices. This is like having a window in your office through which you can see and speak to another person at any time. In some offices a red light flashes on to indicate when another person wants to speak to the occupant. Some systems employ conventions for "knocking on the video door" or give some kind of auditory signal when someone wants to peek into an office by way of a video connection—analogous to knocking on a physical door and asking "Got a minute?" Crucial design issues for these systems center around how people are informed that they or their information are being seen or heard and empowering them to control who has access to the information. One of the design problems of essentially all these experimental systems is that there is no secure way to know when information is being recorded or to ascertain, ultimately, who does or could have access to it.

Cameras and recording devices in public areas of Xerox PARC have been received differently by different research groups. People in one laboratory showed very little concern about the cameras, while in another lab at the other end of the same building resistance was so strong cameras had to be taken out or were never installed. At Apple Computer, a survey of customers at a coffee bar periodically scanned and recorded by video revealed a small but significant proportion of customers who were unhappy to learn that they were on camera. In situations like these, subjects cannot tell who is watching them or what the watcher may do with the information.

In normal physical space, we receive cues and feedback about how we should behave from the environment. For example, if we talk too loud in a library or a place of worship, people may signal us to be quiet. As social beings we exercise control over the way we present themselves in public places. In private, we may confide something to a friend that we would not say in public. People in the coffee bar may be less inclined to engage in casual conversation if they think the boss is watching or if there have been recent complaints about work efficiency.

To test the video edge in a public setting, Steve Mann carried a videocam into a number of department stores in the late 1990s. At the time, Mann was a graduate student at the MIT Media Lab working on wearable computing (see chapter 10). In addition to the videocam, he had a second, miniature camera built inconspicuously into his sunglasses. His practice was to walk up to a clerk in a store, point to a surveillance camera in the ceiling, and ask, "What's that?" Sometimes the clerk would plead ignorance of its function or would say it was part of the temperature-control system. Sometimes Mann was sent to the customer relations department for the answer. He was told several times that the camera was for his protection and that he had nothing to fear if he was doing nothing wrong. He would then lift the large video camera and begin pointedly recording the sales clerk, whose immediate shock and discomfort illustrated graphically the fact that people do not like being filmed without their permission.

Mann's use of personal and wearable video technology was motivated in part by a desire to redress a perceived imbalance (Mann 1996). Video cameras camouflaged by dark-domed "smart ceilings" are now common in stores, banks, and restaurants. In language suggestive of a surveillance arms race, Mann suggests that wearing personal cameras can even the odds; instead of "my gun is bigger than your gun," we have "my camera is bigger than your camera." The security potentially resulting from this arms race could be in the form of alternative evidence. Because video records can be so easily falsified, individuals challenged by a video record might need to present their own video-recorded account of an event. In this vein, Mann imagines shoppers with their own cameras keeping a video record of transactions to counter possibly false claims in a dishonest environment. For Mann, the issue is not so much that surveillance is used as that the balance of power in institutional surveillance is stacked against the individual.

Of course, surveillance information can be recorded with technologies other than video cameras. One such technology, the cellular telephone, must periodically signal its position to the cellular network in order to receive calls. This means that the network has to keep track of all the positions of all the telephones that are turned on in its region. In December 1997, the *Sonntags Zeitung* reported that the Swiss police had been secretly tracking the whereabouts of mobile phone users with records from Swisscom . Instead of discarding the stale data, the network had apparently saved records showing the movements of more than a million phone users. This practice was ultimately declared illegal by the Swiss Privacy Commission and the Swiss parliament.

Is such routine recording of images a violation of privacy? Should privacy laws inhibit large-scale record keeping about personal movements? In this regard, it is interesting to look at the provisions of the broad European directive, 95/46/EC, which grapples with this question. According to paragraph (14), it applies to "techniques used to capture, transmit, manipulate, record, store or communicate sound and image data relating to natural persons" However, subsequent paragraphs spell out exceptions, such as when the video monitoring is for public security, defense, national security, or journalistic purposes. Thus, the writers of this directive did not find the answer to be black and white. Consent, which is a key part of privacy protection, is generally not practical in public places where streams of people are constantly passing by. The directive therefore recognizes certain overriding considerations and in some cases authorizes public surveillance without that consent.

Privacy-Enhancing Technologies

Sure, you can tuck away a few loose threads of information about where you go on the Web by surfing anonymously and using re-mailers to hide where your e-mail comes from, but the legal shield that once surrounded your real-world personal records is now as brittle as a bird's egg.

Charles Pappas, "To Surf and Protect"

During the early 1980s, when artificial intelligence (AI) was receiving a great deal of commercial and media attention, confusion about the field came from naive enthusiasts as well as pessimists. Pessimists tended to demand perfection, which may have discouraged researchers with simple ideas from pursuing them through a potentially long road of incremental improvement. Enthusiasts, on the other hand, suffer from a different illusion: they believe that computers are naturally endowed with superhuman capacities. You can understand these twin dangers by imagining your favorite complex real-world puzzle, such as "optimally timing all the traffic lights in a big city" or "perfectly diagnosing a patient's disease." The silliest and most unrealistic enthusiast would argue that because the problem is difficult and has baffled experts for a long time, the only approach with any hope for success is using artificial intelligence—typically, a computer with a knowledge-based system. The silliest pessimist would continue to focus on the most difficult cases without noticing that most common cases are often simple and straightforward.

However, understanding the confusing relationship between privacy and technology brings out the same extremes. The complexity of the privacy problem is illustrated by the various legislative attempts to address it. There may, admittedly, be no perfect solution. It certainly seems that no fixed set of privacy rules can cover all situations, because the situational requirements and human sensitivities vary so much. On the other hand, even though computers aggravate the privacy problem by making it easier to collect and correlate data about us, they may also provide us with leverage for alleviating the problem. At the very least, exploring the possibilities of using computers to enhance privacy can help illuminate the nature of the requirements.

Key Recovery and Key Escrow

One such form of technology many government agencies and on-line service providers rely on is encryption. *Encryption* is a process of encoding a digital work by using a secret key to render it unusable by anyone without the key. Decoding a work to restore it to usable form is called *decryption*. The preferred method of encoding is *public-key encryption,* in which there are two keys: a published public key and a secret private key. When the private key is used to encrypt a work, the public key can be used to decrypt it.

Encryption is at the core of many systems for secure communications and electronic commerce. Encryption can be used to protect works or documents stored on a file system. Even if someone makes an unauthorized copy of a record, it will not be readable without the key. Encryption can also be used to protect a work on a communication channel (e.g., the

Internet), so defeating the activities of a wire tapper or hacker. In addition, publishers can use encryption to digitally "sign" (identify) a digital work. When a person who has a published and well-known public key encrypts a message with his or her secret private key, the recipient can be sure—because the private key is kept secret—that the message actually comes from the public key holder. Without the private key, nobody else could have encrypted it.

There are two basic ways to defeat encryption: cracking the code and stealing the private key. The security of cryptographic methods thus depends on the mathematical difficulty of the code and the obstacles to obtaining the key through some other covert means, including a computer attack. In a properly designed key, the difficulty of cracking a code mathematically increases with its length.

Throughout the 1980s and 1990s, there was a debate in the United States about the regulation of encryption technology. The various turns and twists in this story are too complex to review here. However, the debate turned (and still does) on the tension between values: national security and law enforcement on one side and personal privacy on the other. The commonly accepted national security argument against allowing free access to encryption technology is that it would let enemy agents transmit secret messages; the use of very long keys would make it highly unlikely that intelligence organizations would be able to intercept such messages and break the codes. On the domestic front, encryption technology would allow organized criminals to cipher their communications, transmit funds electronically, and easily defeat the surveillance capabilities of the civil authorities. Even so, as some advocates of privacy and personal security point out, the availability of commercial products supporting long encryption keys in other parts of the world (including Europe and Japan), renders U.S. laws governing key length essentially ineffective.

Nonetheless, during the early 1990s, the U.S. debate focused on regulating key length to control encryption technology. Privacy advocates argued for keys longer than the forty-bit and fifty-six-bit keys the regulations allow on particular kinds of systems. The routine breaking of forty-bit keys by student programmers using idle computer time have demonstrated that forty-bit keys are vulnerable. Even the fifty-six-bit length used in the popular digital encryption standard (DES) becomes more vulnerable every year as computer speeds increase in accordance with Moore's Law. Recognizing that technology does not stand still, we would need to add about fourteen bits to key length for every twenty years of desired protection.

By the mid- to late–1990s, however, much of the debate had shifted away from the regulation of key length to the pros and cons of *key recovery, key escrow,* and trusted third-party encryption systems. These approaches differ in their details. What they have in common is the understanding that although long secret keys can protect data, they must also provide a way for authorized parties to gain access to secret keys under certain circumstances.

Early key-escrow systems depended on a government department that held in escrow a master key to each encryption device; the key was released to a law enforcement agency as needed and authorized. The term *key recovery* was originally used to refer to systems in which a key is broken into parts and stored on different systems; the key can then be recovered by a process that brings the parts together. The term is now used as a generic name for a variety of techniques.

The value of key recovery goes beyond law enforcement. In the absence of such an approach, owners of data can lose access to their own data if a key is lost or stolen. An organization could be prevented from using the data in its systems when an employee leaves without notice or dies suddenly. Thus, for many commercial applications, there is an incentive to keep secure backup copies of keys.

Much of the debate about key-recovery systems has related to the need to recover keys used for different purposes. For example, it makes little commercial sense to have backup copies of keys used simply to sign messages. A date at which a new key becomes valid can easily be established; after that an old lost or stolen key is no longer valid. Furthermore a key used for a single communication session, such as one telephone call, is not usually worth backing up, as such keys are generally negotiated by encryption equipment at the beginning of each call.

It has been suggested that, from the perspective of law enforcement, it makes sense to be able to recover a key without the knowledge of the user, so that data contained in a system can be accessed during an ongoing investigation. Access to the key should also be rapid—within two hours under some proposed regulations—and obtainable at any time of the day. Finally, all sorts of keys—whether for signing, for communication, or for access to stored data—should be available to authorities.

The operational details needed to organize a national key-recovery system are many. In the United States alone, there are over seventeen thousand local, state, and federal law enforcement agencies. The possibility of insider or corrupt action within any agency, the potentially large number of keys needed, and the desire for rapid delivery of keys to law enforcement agencies present enormous operational challenges as well as the risk of substantially lowering the security of encryption systems in general. At the time of this writing, a technical advisory committee staffed by representatives from different sectors of the information-processing industry is working with the National Institute of Standards and Technology to develop recommendations for a Federal Information Processing Standard for key recovery.

Encryption technology is a central element of all modern approaches to security and privacy. In the United States, the ongoing debate about the need and practicality of regulating encryption devices seeks to balance requirements for national security, law enforcement, commercial viability, and personal privacy. Encryption technology is not, however, the whole story but only a part of what we will need in order to build trusted systems.

Privacy Preferences and Trusted Systems

In chapters 3 and 4, we explored the potential for trusted systems in digital publishing, emphasizing the value of security to publishers. That discussion focuses on the design of systems to guard against unauthorized use of digital information owned by authors and publishers. There is also—from a privacy perspective—a reverse kind of danger relating to consumers' on-line reading or viewing habits. Could trusted systems that safeguard publishers' information also be designed to guard the privacy of consumers?

One step in this direction is the Open Profiling Standard (OPS) proposed in early 1997 by Netscape Communications, Firefly Network, Inc., and Verisign, Inc. Among other things, OPS is designed to respond to concerns about the information web sites gather about their users. In the late 1990s, web site managers interacting through protocols with browsers could employ numerous means of collecting data: they could ask users to fill in forms; they could record the data trail—known as "cookies"—that users' activities generate; they could gather information from the browser about other sites their users have visited; and they could run programs that search individual users' files.

The OPS architecture reflects an interesting blend of commercialism and concern for privacy. The philosophy behind it centers around three principles: (1) control by source, (2) informed consent, and (3) appropriate-value exchange. The control-by-source principle says that parties responsible for creating information should control its dissemination. These parties include both individual users and any entity that gathers profile data. Informed consent means that parties requesting access to a user's profile must receive the consent of the sources before collecting and using the information; they must also provide complete information about how the data will be used. The principle of appropriate-value exchange means that no party should collect information about a person without offering that individual something of value in exchange. Thus, there should be some benefit to the user for providing a profile. If, for example, a party requests a user's e-mail address, the value exchanged might be free updates or news delivered by e-mail.

At about the same time OPS was being formulated, the Internet Privacy Working Group (IPWG) began to develop what it called the Platform for Privacy Preferences (P3P). P3P extends an earlier proposal, the Platform for Internet Content Selection (PICS), adding capabilities for notifying users and obtaining consent to negotiate with them about preferences, policies, and information exchange. In addition, P3P incorporates a machine-readable grammar to facilitate network negotiations. In mid–1997, the OPS and P3P projects were unified.

Although OPS principles fall short of those established by European privacy laws, EU directives, and the proposals of privacy advocates, they head in the same direction. For example, the principles of 95/46/EC, which describes particular individual rights to privacy, are somewhat comparable to the OPS principle that vests control over data in the source. However, the EU directive (unlike OPS) also spells out certain exceptional cases in which requirements are less stringent; for example, when data are collected for national security, statistical, or journalistic purposes. It also specifies institutional requirements for oversight authority and enforcement that are missing from the OPS principles. U.S. law too goes beyond OPS by categorizing certain kinds of data—such as credit information, telephone records, and educational records—as sensitive. In contrast, OPS principles leave these distinctions mainly in the hands of the consumer and other sources. On the other hand, the OPS specification allows users to name a trusted third party that can certify an agent to handle privacy data. If the third party has a reputation for holding privacy values consistent with those of the user, and is competent in monitoring the agent's activities and enforcing sanctions, users may decide that this is as much assurance as they need.

An interesting feature of the OPS approach is that it is intended to be mostly automatic. It would give users full control of data collected about them by managers of a web site. Leaving aside for the moment questions about exactly what rules ought to govern personal data, a crucial question is whether computer systems can be trusted to follow the user's instructions. This point brings us back again to issues about the design of trusted systems.

Threat Analysis and P3P

The threat analysis of trusted-system architectures we discussed in chapter 3 considered both hardware and software attacks. In a 1990s conventionally designed personal computer, viruses and rogue software are serious security risks, because they can make arbitrary changes to the files and software on a system. Instead of radically redesigning the hardware of the installed base of systems, the approach we considered controls the loading of all programs—checking that they have been certified and verifying that they have not been tampered with. Any approach with a lower level of security than this is, in principle, subject to arbitrary attacks by unrecognized viruses.

This is not to say that OPS or P3P efforts, as drafted in late 1997, were unaware of trust and security issues. In fact, P3P's developers intended to make it possible for web sites to express their privacy practices explicitly and for users to make choices about those practices and have their computational agents enforce those choices. Indeed, the draft documents outlining the architecture and grammar for P3P even mention a "trust engine," although it is defined only as a mechanism for evaluating incoming statements before a decision is made—typically just some code associated with a network browser. It is not specified as a trusted system in the sense we use the term in chapter 3. The idea of using digital certificates to certify that sites actually follow the practices they express is undeveloped in P3P specifications (at least as of 1997). Also undetermined are the mechanisms used to ensure that computers participating in the dialogue about practices and preferences have not themselves been compromised or tampered with in some way. In short, P3P has no solid provisions for protecting personal data from the kinds of threats security analysts consider most important.

This brings us to a vulnerability of OPS as well. In the absence of a trusted-system architecture, a rogue program or virus could, in principle, render ineffective all of the user's specifications about how he wants his data protected. In other words, a virus entering the system at any time could change the user's profile or change the way programs see the profile so as to make the protected data available to unauthorized systems. It could also cause the system to provide web sites with false or bizarre profile data about the user and his or her interests or habits, with consequences that could last for years.

Like systems for copyright protection, the trust issue for OPS and P3P extends to safeguarding the information after it has been released. The web site itself must be a trusted system that will tag the user's data with instructions about its use and keep that tag associated with the data as long as required. In addition, it should provide the tagged data only to other trusted data processors upholding the same rules. In this regard the privacy rights for personal data are essentially identical to usage rights pertaining to digital published works. The analogous infrastructure would require all systems processing users' private data to be the kind of trusted systems we described for copyrighted works. By building systems that benefit both consumers concerned about privacy and publishers bent on protecting copyrights, trusted-system vendors could greatly increase the perceived value of their products.

The central focus of debate about privacy on the Internet is the handling of personal on-line data. The basic goal of P3P and OPS, to bring the users' private data under their control and to put the responsibility for handling the data under the automatic control of the computer systems, has not yet been achieved. In all fairness, we should point out that these architectures were proposed as something that could be done right away, without first creating a trusted system infrastructure. These proposals bring us, therefore, collectively face to face with an interesting point at the Internet edge, where the advantages of networking and a desire for privacy and security create both risks and opportunities. Building an infrastructure to bring reliable balance and security will not be a one-off, quick collaboration. It will require more time and a deeper investment.

Visibility, Accountability, and Scale

Enthusiasts of the Internet often rejoice that it has created a virtual "global village" replete with opportunities for new and cozy relationships with people around the world. Yet, if we compare the current situation with regard to trust and computers to the development of trust among people in a small village, we find three elements that impede the creation of on-line trust.

- *Visibility.* Much of what happens on a computer is just plain invisible. When data are collected by a computer or sent to a third party over a wire, the activity is hidden from the data subject.
- *Accountability.* When data travels from one computer to another, or is combined with other data, information about how the data should or should not be used is not included. Data carry no tags identifying their source or specifying what agreements users have made for their use. If people or computers use data in ways that violate such agreements, there is no automatic accountability.
- *Scale.* The social methods that work to create cohesion and trust in a small village depend on people knowing each other. The methods that work for states and nations depend on legislative techniques that also require that people be identifiable and that they reside within a fairly restricted geographical area or legal jurisdiction. The Net, by contrast, connects people and organizations that may be hard to identify and that live in all parts of the world.

These elements are central to the fears many people express about on-line privacy. Furthermore, any approach to data protection that fails to address these differences will face continued resistance to adoption of a new technology. Without visibility, people can fear that things are happening that they cannot see. Without accountability, there is a sense that releasing data is like letting the genie out of the bottle. Nobody knows where the data goes, and it becomes difficult to hold people or organizations accountable for their use of them. The issue of scale makes it necessary to match the methods for handling private data to the scale of the problem. If the problem spans computers across networks, then the solution needs to operate in the same space.

These three factors may seem to make the data-protection problem totally intractable. For example, we might think that invisibility and intangibility are inherent to efficient computers and communication. Basic physics tells us that something in a computer has to move very fast in order to compute quickly and to travel long distances quickly. The smaller and lighter things are, the more quickly they can move. Today's computers use electrons, which travel at about half the speed of light and are far too small for computer users to observe directly. Even sound, which seems almost instantaneous when we are talking with someone in the same room, is immensely slower than the electrons that carry our voices over telephone lines. If telephones used sound waves instead of electrons for coast-to-coast calls in the United States, conversation would be very stilted, because it would take over three hours for the sound of our voices to travel one-way across the country. We would need to wait at least six hours for the answer to any question. Thus, smallness, lightness, and intangibility are fundamentally important to our ability to interact with each other over national and global distances.

However, despite these observations, computers can be among the most powerful instruments for making things visible. In the past few years, research on information graphics has grown enormously, bringing new methods of visualization to both scientific and information-management applications. Computers can pick out tiny events from massive amounts of data and highlight them. In short, lack of visibility is *not* an inherent property of computers.

We might think, too, that lack of accountability is intrinsic to information processing. After all, data can be combined from many sources. Data files can be merged. When data appear in a program or computer memory, who can say where it came from? How could any person or any computer be held accountable? Nonetheless, computers are actually among the most powerful instruments for accountability. They are the crucial foundation of record keeping and accounting that takes place in the financial houses of the world and the accounting departments of all small businesses and major corporations. They are the most reliable memory for the detailed personal budgets of most owners of personal computers. In short, lack of accountability is *not* an inherent property of computers.

Finally, we might think that scale as well is innate to computers and will inevitably defeat any system of privacy. As suggested above, there is an interesting parallel between copyright and privacy. Copyright law governs the use of published works just as data-protection laws in privacy govern the use of personal data. When the digital medium and desktop publishing first appeared, publishers were afraid that unregulated copying on computers would undercut their ability to make a living. They withheld their works from the digital medium. With the network becoming ubiquitous, citizens concerned about privacy are experiencing a similar uneasiness. They fear that copying and the unregulated use of their personal data will undermine their privacy and interfere with the enjoyment and independence of their lives. They are inclined to withhold personal data from the Net—even though they are attracted to the advantages the Internet could provide if they did not need to worry about privacy.

Their concern brings us back to the idea of creating trusted systems for personal data. In such systems, privacy rights for personal data could be tagged the same way we can tag usage rights on published works. Analogous benefits would accrue to owners of data. The same security issues for making computers systems worthy of trust would have to be met, and similar institutional and infrastructural operational problems would have to be solved. The legal basis for our proposed trusted system for digital publishing could be established by shifting the basis of protection from copyright law to contract law. In trusted systems for privacy, the legal basis for protection could also shift, from privacy law to contract law. Privacy rights could be treated as digital contracts for using personal data and enforced by trusted systems.

Privacy Rights and Trusted Systems

Today's computers and installed software are not programmed to honor privacy rights. As suggested in the discussion of OPS and P3P, one method of ensuring privacy with trusted systems is to extend the notion of compliance and trustworthiness to other systems in the chain. Billing systems, for example, could be tested for compliance with policies for data accuracy, reporting, and aggregation. If a consumer's personal computer or PDR is a trusted system and the interacting financial clearinghouse is not, the latter would be unable to pass the challenge-response protocol for processing billing and usage data and would not gain access to personal data. The user's trusted system would only communicate data to a financial clearinghouse or data aggregator that is appropriately certified.

In digital publishing, as discussed in chapter 3, exceptions to allow free use of a digital work would be governed by the fair use doctrine. Analogous conditions of free access to personal data could be limited to rare and suitably qualified and regulated interventions in the public interest, such as law enforcement, national defense, and (possibly) census taking. For example, government investigators working on a criminal or other official case might want access to private records contained in a trusted system in order to monitor an individual's reading habits or an organization's publications and subscribers. Depending on the surrounding facts and circumstances, such access could be a legitimate use of authority or a violation of civil liberties. Presumably, the protections of the Fourth Amendment against unreasonable search and seizure would apply to trusted systems. Would present-day laws—such as statutes prohibiting unauthorized wiretapping—adequately safeguard the privacy of users of trusted systems while taking the legitimate interests of government into account? Or would we need new protective legislation specifically tailored to trusted systems? These are important questions that we must leave for another day.

The trusted-system approach to protecting and accessing personal data would help balance commercial interests with privacy interests. However, one consequence of introducing the privacy issue into the overall scheme for protection of intellectual property is that it adds financial clearinghouses to the set of participants whose interests must be balanced. The financial cost of building trusted systems is substantial. In their absence, initiatives like OPS and P3P may be workable first steps toward a privacy infrastructure. Financial clearinghouses for usage data have already been built for applications such as DirecTV subscriptions, suggesting that there is a financial incentive to develop such services. Privacy concerns argue in favor of going the extra step to create software systems that would comply with codes of trustworthiness according to specified privacy rights. Stakeholders in this area would include financial clearinghouses and data aggregators, because their systems would need to keep track of and honor privacy rights.

Trusted Systems in Untrusted Institutions

Ultimately, however, even trusted systems in institutions staffed by fallible (and perhaps untrustworthy) individuals will not be able to protect privacy. The pragmatic aspects of protecting private data are more difficult than those of protecting copyrighted works. A copyrighted multimedia encyclopedia on a CD-ROM contains almost a gigabyte of information. Even shorter copyrighted works often contain tens of thousands of bytes of data. Because it is quite expensive to recreate such works by hand, trusted-system assurances that a work cannot be casually copied from one machine to another will probably be borne out.

In the case of private personal information, however, the amount of information to be protected is often much less, and this creates a different risk. A lot of sensitive information—such as family names, social security numbers, salaries, credit card numbers, birth dates, addresses, and medical records—can be packed into very little space. Suppose that a computer system inhibited the copying of such information to an untrusted repository but allowed a clerk to display it on a computer screen. It would take only a few keystrokes for him or her to enter the information on a second computer, with no safeguards beyond the trustworthiness of the person at the keyboard. This is the "re-keying issue." Once private information is removed from the trusted environment, it can be sent anywhere.

This scenario suggests that control over individual information might need to be much tighter than corresponding controls of copyrighted works. One approach to the re-keying problem might be to allow display of personal data only as part of a suitable aggregation; the trusted system would therefore refuse to display information about a particular person but would provide data about groups so large that sensitive personal information about any individual would be difficult, if not impossible, to discern. Another measure would be to periodically create trick records, that is, records of nonexistent people, then offer this information to suspected untrustworthy sources and watch to see whether the data are leaked.

Finally, proactive ways of checking the privacy performance of organizations on the Net may be crucial to maintaining privacy. Trusted systems, by themselves, will not be able to ensure that small amounts of private data released only to trusted organizations will stay private. In this regard, the old advice never to divulge truly sensitive information still seems sound.

Privacy Services on the Net

In the absence of trusted systems and widespread adoption of privacy principles, a few privacy-related services appeared on the Net in the late 1990s. These included anonymous e-mail, anonymous web surfing, and privacy-related search services. There are different notions of anonymous e-mail, but the most basic kind is a service that will remail an e-mail message without any identifying information as to its source. Anonymous web surfing is a service for browsing that shields information about a user from protocols that attempt to read data off the individual's computer. Finally, subscribers to services such as Privacy Inc. can search through three hundred state, local, and law enforcement web sites to learn whether their names are listed on any of them. Yet other companies are providing services to tell a subscriber when anyone requests his or her credit information.

Although these services are intended to advance the cause of privacy, their effectiveness is open to question. Anonymous web surfing may prevent the linking of personal data for free browsing. In principle, it could be combined with anonymous digital cash for privacy in for-fee browsing as well. Anonymous e-mail raises questions analogous to those raised about caller-ID services in the telephone system (i.e., about a caller's true identity). Neither of these approaches has yet gained widespread use.

Reflections

> We used to live in villages. We are moving toward a global village now, so perhaps it's inevitable . . . we will know all about our neighbors again.
> Carole Lane, *Naked in Cyberspace*

When Steve Mann asked about the need for surveillance cameras in a department store, the clerk's response was like a line from a bad movie: "You don't have anything to worry about if you haven't done anything wrong, do you?" Similar questions come up when people discuss anonymous e-mail ("What are you hiding from?") or anonymous web searching ("What are you looking at that you don't want people to know about?"). However, in a social context, trust comes from experience with other people. Our concerns about privacy and computers arise from the need to trust people we do not know and whose motives are unclear to us, people whom we often cannot see and whom we will probably never identify.

The mechanisms we use to create conditions of order and trust in social groups vary with the size of the group. The larger the group the more formal are the structures and mechanisms—such as codified laws and courts—that regulate our relationships. In the area of privacy, creating specific and practical rules has proven to be a difficult undertaking.

Part of the difficulty we have in defining such rules is that the value of privacy competes with many other social values. In the middle of a criminal investigation, personal privacy can compete with effective law enforcement. A video surveillance camera in an isolated parking lot pits our desire for privacy against our need for personal safety. When someone is gathering facts for a news story, privacy competes with freedom of information. When someone uses personal data to target potential customers by focusing direct mail advertisements on their interests, privacy competes with business efficiency. When privacy regulations affect the flow of information across national borders, privacy may compete with international trade. When a child retreats to his room to dance without anyone watching and evaluating him, privacy and individuality compete with socialization and conformity.

Thus concerns about privacy interact in complex ways with many aspects of our social lives. When these values come into competition in real life situations, the appropriate point of balance depends on the particulars of the situation. Imagine, for example, that computer processing of video images has improved so much that it is possible for computers to routinely recognize faces and identify people automatically. This is not yet practical in the late 1990s, although, given some extension of the present capabilities of technology, it is no longer utterly unimaginable. Suppose that Joe is walking down the street on a summer day near a department store. A street camera picks out his image and forwards it to a computer network for "customer opportunity" processing. It checks the quality and style of Joe's clothing, his credit rating, his cellular telephone number, and a few other items. Moments later his cell phone rings, and a salesperson at the department store invites Joe to take advantage of a sale on some coats in his customary style, size, and price range. Is this surveillance, computer-image processing, and automatic use of personal data for very direct marketing a violation of Joe's privacy?

If so, consider that in some street markets, essentially the same kind of interaction takes place when a streetwise sales clerk sizes up Joe as he walks by and makes a pitch. In court cases in which someone claims a violation of privacy, the court's decision has often turned on whether a person in that circumstance can reasonably expect privacy. Perhaps in we should have little expectation of privacy when walking down the street, whether the interruption is computer aided or not. The expectation depends on the culture of the street.

The example illustrates the great complexity of our rules for privacy. Not only does privacy interact with many other social values, but those values themselves depend on culture. Given the wide variations in human culture, it is probably not realistic to expect to set a global standard for something as complex and as subjective as privacy.

Nonetheless, in 1997 the International Standards Organization's (ISO) Technical Management Board looked into the possibility of establishing a privacy standard. The U.S. representatives opposed the idea, arguing that without European commitment to write and use such standards, producing them would waste the ISO's resources. At least as of mid–1998, neither the EU nor the bodies in charge of individual European national standards support the ISO effort.

A View from the Edge

We can look at all the changes that have occurred in privacy regulation over time as evidence that the legal definition of privacy is still evolving. From this perspective, what we need is to get the laws right. Unfortunately, the differences between cultures and the complex trade-offs in competing values that arise in real situations do not bode well for establishing uniform and universal laws of privacy. Furthermore, to be workable any approach will have to address the three obstacles to human acceptance—visibility, accountability, and scale

We can, however, thoughtfully consider a different approach to privacy at the Internet edge. Computers have helped to create new problems of privacy; some of the very elements that make them so attractive for electronic commerce and digital publishing on the Net can also undermine our privacy. Nonetheless, we may find the resource we need to deal with and transform the privacy problem in the very nature of computers themselves. We need to discover a way to make invisible things visible, ensure accountability, and scale up to a global network.

In this chapter, I have made the point that although we don't usually associate "publishing works" with "keeping data private," putting personal data on a Internet-connected computer is akin to publishing it. Thus we can also speculate that the form of computer best suited to digital publishing, and the usage rights best adapted to protecting published works, will be analogous to those needed to ensure privacy rights for personal data. Just as usage rights and trusted systems can tame computers that are engines of opportunistic copyright infringement, privacy rights and trusted systems can prevent computers from being engines of opportunistic intrusion and privacy violation. We may find that building trust into our computers will enable us to offer a new kind of trust to people we can't see—people we don't watch on our computers but who share the world with us.

9

Strangers in the Net: Access, Diversity, and Borders

If you give people a chance, they are less interested in turning the Net into a world forum than a back-yard fence. There's a new Tower of Babel, and it has an Intel sticker on the side.

Geoff Nunberg, on National Public Radio's *Fresh Air*

What do you mean saying that the world is not a global village? I am from a village in India. People from my village go off and marry someone in England or the U.S. or Africa. The world is a global village. It's the strong networking which sets the tone for easier, smoother global communication. I can talk to someone in England—by e-mail or phone or anything else—easier than I talk with my neighbor.

Mali Sarpangal, 1997

How is the worldwide connectivity of the Net changing our lives? I found myself returning to the question again and again in the course of writing this book. In several chapters I have addressed particular fragments of an answer—the potential for a worldwide marketplace in digital goods, devices to make the network more portable and ubiquitous in our daily lives, methods to help us make sense of the overabundance of information, ways to preserve our more and more elusive personal privacy. I've found ample evidence of national and international trends on which to base these fragments of an answer.

Still, I felt a nagging sense that my picture of the emerging Net was incomplete in important ways. In particular, I was concerned that my life in Silicon Valley may be blinding me to important phenomena or overfueling my enthusiasm for the Net. In this part of California, the source of so much computer technology and culture, the Net has a sheen of unreality about it. Here the stuff of somebody's technological dream shows up quickly in the local electronics stores and later spreads throughout the world. The same could be said of only a few other areas of the United States and the world. Has this constant reinforcement of our technological dreams led me to an exaggerated belief in the importance of the Net that doesn't reflect the reality emerging elsewhere?

The world, after all, is not the Silicon Valley, and the people of the world are not just like the people who live and work here. In this chapter, I try to step outside the Valley—to different parts of the United States and the world beyond—to ask the question again, seeking answers in the contexts and lives of people far removed from the technological centers. I make no claim that my exploration is complete, and I realize that it raises many new questions. The contrasts between the various natural and cultural contexts are extreme enough to lead, inevitably, to very different responses to the Net; they reflect as well the diverse expectations that the world's peoples have about the Net and how it will affect their lives.

No Phones, No Electricity

In 1998 I spent the month of June trekking around Nepal and western Tibet with a group of Europeans. This journey raised many questions for me about the influence of the Net in third world countries.

Telephones and electricity—essential infrastructure for Internet communication—are rare and work only intermittently in western Tibet. In a guest house we stayed at in Darchen—a small town visited by many travelers to Mount Kailash—the lights came on dimly at 9:00 P.M. and went off at 10:30 P.M. There was no way to plug in an electronic device during the day. While the electricity was on, there was no way to turn the lights off, except by removing the light bulb. On the frequent nights when the electricity did not work at all, the proprietor of the guest house brought candles to the rooms. The guest house, we were told, had the most reliable electricity in town.

Other factors not directly related to the electricity supply also affected the local context for the Net. The ground floors of the guest house, and every other establishment in Darchen, were dirt. There was no plumbing of any kind. A nearby creek was used by animals and for clothes washing and for cooking water. You boil and treat water before drinking it. Western Tibet has practically no paved roads. Westerners travel over long distances chiefly by Toyota land cruiser or truck, driving over faint trails in the high desert or along river beds. Locals walk a lot. About 20 percent of the population are nomads who tend their yaks and mountain sheep in the high desert country.

Telephone service is even more rare than electrical service. Larger towns, especially those with a governmental presence, often have a microwave tower that supports telephone communication. You go to a dedicated telephone building and wait until one of the two or three phone booths is available. After making your call, you pay cash to a person at a teller's window. Phone service is intermittent, too, because it depends on the availability of electricity.

In the next few years the planned satellite telephone system will provide greater connectivity for Western travelers and relatively wealthy Tibetans, even when they are trekking in the mountains or driving on the high desert. However, the need for generating equipment to keep batteries charged and the relatively high cost of connection are likely to keep the routine use of telephones beyond the reach of most residents for many years.

One might expect from my description that the Net is completely unknown to most people in the region, but that would overstate the case. The majority of trekking expeditions to western Tibet begin in Nepal, often in Kathmandu, a third world city with an international airport and headquarters for numerous trek organizers. Even though the electricity in many portions of Kathmandu goes off for several hours a day, one can hardly walk down the street in the city without seeing an advertisement for e-mail. Some restaurants feature "cyber corners" with computers set up for Internet access, and many trekking companies have their own web pages. The business card of a shaman that our group visited there lists an e-mail address for her earthly communications.

In short, the Net is now present on the fringes of Tibet in the large cities, competing with the fax machine as a useful technology for international communication. Tibetans are also becoming acquainted with television—the World Cup soccer matches dominated the programming we saw in shops and restaurants. Television advertisements mostly feature home appliances like refrigerators, but they also mention computers. Messages from the outside world already come in via television and the occasional solar-powered radio. The Net is coming to Tibet too, but at a much slower pace. In the late 1990s its growth was greatly limited by the weak economy and government regulation.

The Net and Cultural Diversity

The year before my trekking holiday in Tibet, I took another extended trip, this time to Italy. My earlier book, *Internet Dreams: Archetypes, Myths, and Metaphors* (1996), had just been translated into Italian, which gave me the opportunity to spend some time talking about the Internet with members of the Italian press.

One question that came up in these conversations was whether the Internet was a pipeline for spreading American culture to other parts of the world. There was some sensitivity about this issue as certain highly visible American products, such as McDonald's restaurants and American movies, had already made inroads into the Italian market. The journalists readily agreed that these American imports would not be succeeding as well as they are if Italians did not have an "appetite" for them. I observed that, in any case, people can put anything they want on the Net—including propositions for Net standards—and that viewers and consumers can decide for themselves what they want. Hearing this, one journalist quipped, "It's the Wild West all over again. How American!" This comment stuck with me. It suggested that I was so enmeshed in the Silicon Valley that I sometimes did not even notice the elements of culture (e.g., the "Wild West attitude") that permeate my own thinking.

To many people, unwanted foreign products or influences are akin to species pollution. In many parts of the world, local ecologies have been radically changed—sometimes in just a few years—by the introduction of foreign species that spread without check. In some coastal and hilly areas of California, South American pampas grass and broom from Scotland have overrun native plants. Fishing of native species in many bays around the world has been wiped out by foreign organisms introduced when foreign ships emptied their ballast tanks while in harbor. An exotic Pacific seaweed, *Caulerpa taxifolia*, escaped from a German aquarium two decades ago; it is now spreading throughout the Mediterranean from the southern coast of France. Where it spreads, the toxic plant wipes out sea anemones, starfish, crabs, shrimp, and other animal species and suffocates any native plant in its path. In Louisiana, swamplands have been overrun by the nutria, a small mammal whose population has soared to over twenty million since a hundred and fifty animals escaped from a fur farm earlier in the decade. Extensive changes in Australia's ecology began many years ago when Europeans explorers and settlers introduced dogs, sheep, and other mammals.

Such damage to native microbial, plant, bird, and insect populations abounds, often accompanied by unfortunate consequences for local agriculture. The foreign invaders spread quickly because other species that could keep them in check are missing in the new environment. Furthermore, since any ecology is a complex system, experts are often uncertain about what measures to take to restore balance without causing other, unwanted effects.

The species-pollution metaphor also fits some commercial products very well. During my stay in Venice, I caught sight of only two or three McDonald's restaurants. Yet, in one four-day period I noticed that approximately two out of three pieces of litter in the streets bore McDonald's labels. I have no particular dislike for McDonald's, and I eat there on occasion. I cannot explain why so much of the litter in Venice comes from McDonald's. I suspect that McDonald's uses more disposable wrappers and containers than Italian take-out restaurants and is frequented by teenagers, who are less inclined to dispose of their trash properly. However, other Venetian establishments—such as bakeries and the wonderful gelato shops—also dispense food in paper—and there are many more of them.

In Tibet, too, I saw a dramatic example of pollution caused by foreign products. Canned soft drinks and bottled beer were apparently introduced into western Tibet in a fairly big way during the 1990s. Because there are no effective garbage or recycling services in most of the country, the land around the stores and on the outskirts of small towns is always piled high with a several years' accumulation of discarded bottles and cans. They are not buried or organized in any way. Perhaps at some point it will become worthwhile to "mine" these garbage piles for the glass and aluminum they contain. In the meanwhile, the dumps are very visible and unfortunate reminders of Western influences and their ecological consequences.

The proliferation of foreign biological species and consumer waste products fit the metaphor of ecological pollution quite well. But is it particularly apt for describing the cultural effects of the Internet? Digital goods do not pile up in public spaces the way litter can. Furthermore, the quantity of cultural products on the Internet accessible to the average third world person is still small and much less influential than that conveyed by television, which is itself still relatively rare in many parts of the world.

So what degree of cultural mixing is actually taking place because of the Net? Conversations on this subject with acquaintances in Nepal and Tibet gave me a few partial answers. The Net technology itself—ranging from computers, to e-mail, to the creation and browsing of web pages—is available to the growing numbers of technologically literate people. Although computers and Net communications are not within reach of most people in the third world, those who do use the Net tend to be more influential and entrepreneurial. They employ reasonably up-to-date technology. The hardware and software on display in the stores in Kathmandu are not much different from what is sold in Silicon Valley. It's not as if the stores were carrying ten-year-old computers. There is a time lag—and there are some older computers in use—but that's true in Silicon Valley too.

East Asian newspapers carry stories about the Net. People using the Net tend to be familiar with the politics and geography of the West and to be interested in Western developments. Besides the usual business reasons for studying English, computers are giving young people an additional incentive. Several people told me that learning English was essential for learning about computers because the best training materials are in English. Even where basic instructional materials are available in the local dialect, the Net's rapid pace of change gives those who learn English a decided advantage.

The ecological-pollution metaphor would have weight for thinking about the effects of the Net if we could show that it actually decreases cultural diversity. Does it? Certainly the decline of cultural uniqueness was an issue long before the Net came into being. What is not clear yet is how much the Net contributes to this phenomenon, as compared with other, pre-existing influences. So far, however, what we see is some tangible *increase* in diversity, as local cultures combine with the Net culture and as people form special interest groups—sometimes internationally—to pursue topics in which few of their neighbors are sufficiently interested.

When the Net technology evolves to provide low-cost, high-speed, and portable access to all kinds of digital material—supplanting television, newspapers, and other media in a single unified medium—the impact of the Net on local cultures worldwide will be huge. Many people will carry around portals to the networked world, and centralized information and entertainment services will compete for world market share—and mind share. Such a picture, however, is not the reality of the late 1990s. The Net in the third world is still accessible only from desktops over slow and unreliable telephone lines. There are no good examples of cultural displacement on the Net—in part because so little of culture is on the Net anyway. There are also few examples, so far, of cultural mixing on the Net, or of new Net technologies originating in third world countries. At present the Net principally excels at giving us—Westerners planning a trek in the Himalayas or (say) Nepalese searching for information about computers—relatively cheap access.

A friend of mine recently told me about getting off a plane on a Honduran island for a skin-diving trip. An unknown woman approached him in the airport to say that his colleague had missed a connecting flight and would be a day late. On the island there was an "Internet Cafe" next door to the sole microwave tower on the island. While in the cafe, the woman had received an e-mail asking her to deliver a message to a guy wearing a brown hat who would get off the plane the next morning. Such examples give us a sense of the "small world" the Net can create, but they do not signal an imminent period of rapid change and globalization stemming from the Net. Rather they reflect the niche uses and applications of the Net that are proliferating while the larger issues of economic development are working themselves out in the third world at a slower rate of change.

Computers and Democracy

> Freedom is fostered when the means of communication are dispersed, decentralized, and easily available, as are printing presses or microcomputers. Central control is more likely when the means of communication are concentrated, monopolized, and scarce.
>
> Ithiel De Sola Pool, *Technologies of Freedom*

> As a great social leveler, information technology ranks second only to death. It can raze cultural barriers, overwhelm economic inequalities, even compensate for intellectual disparities. In short, high technology can put unequal human beings on equal footing, and that makes it the most potent democratizing tool ever devised.
>
> Sam Pitroda, *Development, Democracy, and the Village Telephone*

Italy and Tibet, my two major travel destinations during the time I was writing this book, represent extremes. Not only do the two countries differ radically in their levels of economic modernization and use of high technology, they also offer very different opportunities for the exercise of democratic and personal freedoms.

For several years we have heard numerous reports that communications technology is playing a pivotal role in the emergence of democracies worldwide. Well-known accounts include stories about the use of e-mail and fax machines for exchanging news during the break-up of the U.S.S.R in the mid–1990s and during the Tianamen Square riots in China in 1989.

Until recently, however, evidence for a correlation between democracy and communication technology was mostly anecdotal. Does the Net really live up to its touted potential for profound influence on human freedom and government? At first glance, this possibility seems at variance with the commercial focus and niche uses of the Net we have considered so far. Naturally, we would need to ground any conclusions about such a broad question in more than anecdotal accounts and would have to study a wide range of national contexts to mitigate the hazards of biased sampling. The observations and conclusions about this issue I present here draw heavily on a study by Christopher Kedzie (1997).

As Kedzie suggests in his definitive article, democracy is an abstract concept and difficult to measure. As such it has a subjective quality and is susceptible to interpretive bias. Countries are complex. A democratic country may have nondemocratic organizations and practices, and vice versa. Furthermore, democracy is multidimensional. Yet, Kedzie observes, democracy rankings of nations by independent authorities using different criteria agree quite broadly; this suggests that democracy, whatever it is, can be recognized, at least intuitively, by Western analysts.

In Kedzie's study, democracy is the dependent variable. His evaluation of a country's democracy is based principally on the average of two measures: a ranking of political rights and one of civil liberties, as established by Freedom House (a nonprofit, nonpartisan organization that promotes democratic values). Communications technology, the proposed independent variable, is a broad category. Computers can send faxes, and radio and television cables can transmit e-mail. Kedzie decided to focus on e-mail, because it enables people to engage in interactive discourse. He therefore created a metric he called *interconnectivity,* which is based on the number of nodes reported in each country on the four primary e-mail networks: Internet, BITNET, UUCP, and FidoNet. Kedzie's detailed study also takes into account such other variables as wealth, education, and historical factors. He considers the nations of the world as a single group and also according to regional subgroups.

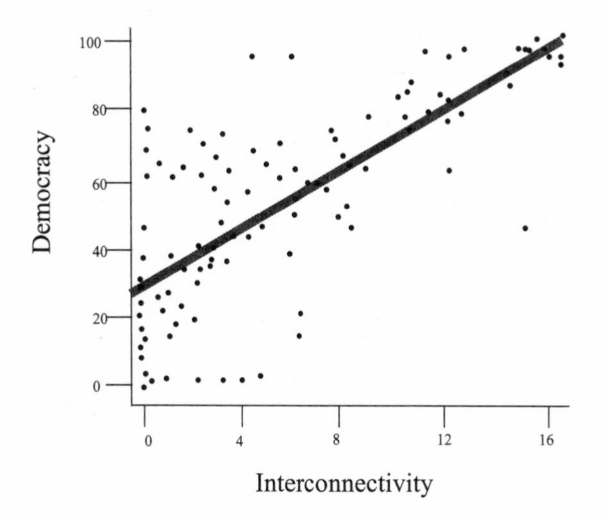

Figure 9.1
Democracy and Interconnectivity. (After Figure 1 in Christopher R. Kedzie, "The Third Waves," p. 113. In *Borders in Cyberspace: Information Policy and the Global Information Infrastructure,* Brian Kahin and Charles Nesson, eds. Cambridge: MIT Press, 1997.)

What is striking about the results of the study is that every analytic perspective, every model, and every set of statistical tests he looked at supports a strong correlation between increased interconnectivity and democracy. Interconnectivity correlated strongly not only with the assessed level of democracy but also with changes in democracy over a ten-year period. The correlations had especially high substantive value and statistical significance in regions where political transformations had been rapid.

Still, Kedzie is cautious about claiming that increases in interconnectivity *cause* a country to shift toward democracy. His studies suggest that level of democracy and level of communication are correlated. As he put it, the most plausible relationship between democracy and networked communication is probably a virtuous circle with positive feedback.

For me, the statistical account was most illuminating. It is easy to hypothesize about how the Net *could* favor democratic thinking and participation. Two-way interactive communication on e-mail is different from the one-way mass broadcast of radio and television; moreover, e-mail communications are not centrally controlled. It is easy to see how e-mail *could* be used politically for one person to speak to many. After all, the potential fan-out of e-mail and web pages is different from the one-to-one communications of the telephone or fax machine, and we know that e-mail is used routinely in this way in special interest groups. But it is one thing to say that such communications are possible and another to say that e-mail really does have a substantial democratizing effect for people in all sorts of countries living in all sorts of cultures and economic conditions. Nonetheless, Kedzie's study may provide an eagle's-eye view of such democratizing effects, one that is broader and more compelling than anecdotal accounts or theoretical perspectives.

Power and Rules for a Global Net

> In spite of the fact that the Internet is a working anarchy, there still has to be someone or something that sets the rules. If that sounds paradoxical, it is. The body that provides the rules is the Internet Society. It's sort of a benevolent uncle to all the Internet users.
>
> Network Business Services web page

> The implications of this new global communication—or interactive society—are quite obvious: national borders are increasingly disappearing within cyberspace; facts, issues and opinions interact anonymously; hypertexts produce new cultural linkages.
>
> Ingrid Volkmer, "Universalism and Particularism"

In the 1960s sociologist and critical theorist Adolf Berle wrote a book with the simple title *Power* (1969). In it, he articulates theories about the principles, nature, and exercise of power in human affairs, substantiating them with examples from world history and corporate life. He observes, for example, that when a given power is not being exercised, people appear spontaneously to take it up. A memorable example of this occurred on the streets of New York City during the 1965 electrical blackouts when citizens stepped into intersections to direct traffic after the traffic lights stopped working.

A key factor legitimating an individual's or a group's exercise of power is the perception of the majority that the power wielder or leader is acting effectively to reduce or prevent chaos. In the electrical blackout, no official appointed the self-styled traffic "cops." No governmental or democratic body sanctioned their work or invoked penalties on those who failed to follow their directions. And yet, driver after driver arriving at intersections, faced with these volunteers in action, chose, for the most part, to follow their directions. In more enduring situations, dictators and others with dubious reputations often remain in power because there is a widely shared, similar belief that the chaos and disorder precipitated by a power vacuum would be even worse.

Do the same kind of issues about power arise in the context of the Internet? What sort of Net activities need governance? What persons or organizations wield power on the Net? How—given that the Net crosses national borders—does governance of the Net relate to the laws and jurisdictions of existing governmental bodies? What are the power dynamics of corporate enterprises operating on the Net?

One of Berle's laws of power is that power is exercised through and depends on institutions. These institutions limit, come to control, and eventually confer or withdraw power. Historically, the Internet grew out of the ARPANet, and its governance was highly influenced by a combination of academic and military interests. The crucial early concern of ARPANet users was establishing technical compatibility so that the different kinds of computers in use at the time could communicate with each other. A great deal of cooperation was needed to accomplish this goal. What evolved from these efforts was a spirit of open review and a system of staged commentary for establishing the standards and protocols for the Net. Since then, the Net has gone through many stages of shifting sponsorship, internationalization, and, inevitably, accommodations with other organizations. Still, the spirit of cooperation that determined early ARPANet standards lives on in an organization known as the Internet Engineering Task Force (IETF). It is an international community of network designers, operators, vendors, and researchers concerned with the evolution of the Internet. Membership is open to any interested individual.

Figure 9.2, based on an Internet web page, shows the IETF as part of the Internet Society and maps the society's principal relationships to other organizations. The Internet Society, which describes itself as a nonprofit, nongovernmental, international, professional membership organization, focuses on standards, education, and policy issues. Like ARPANet in its early days, its core concerns are technical issues about communication standards—such as domain names and Internet addressing. The various

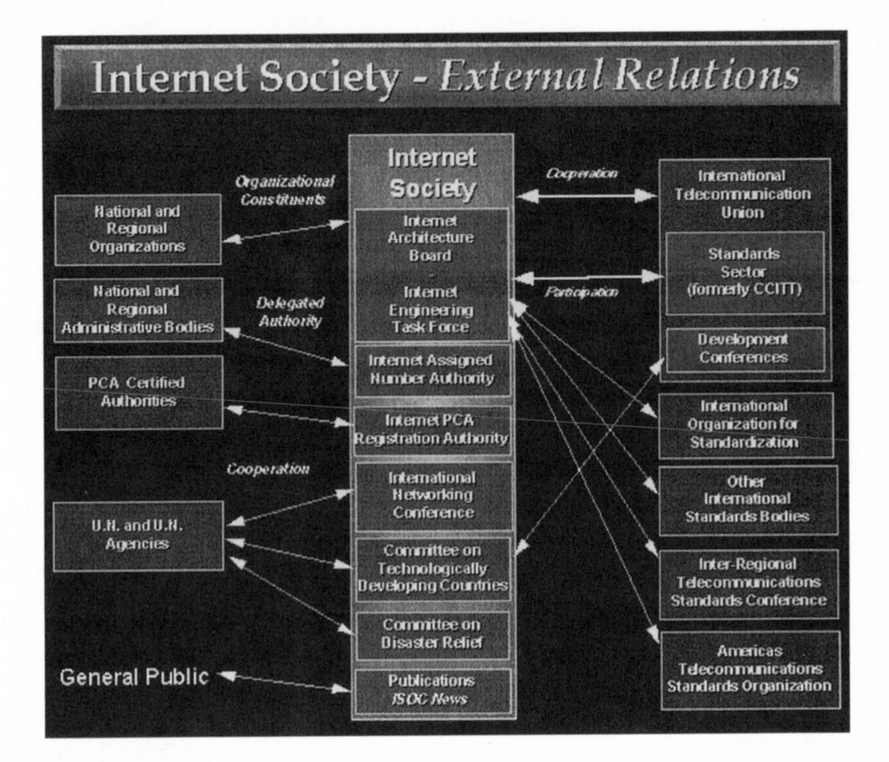

Figure 9.2
External Relations of the Internet Society. (From Dan Larsen's web page: http://www-ccs.ucsd.edu/~dnl/sophe/external.html)

proposals for new standards are available for public review and comment on the IETF's web pages.

There are, besides the Internet Society, many membership organizations concerned with the evolution of the Net. Another influential group is the World Wide Web Consortium (W3C), which was organized in October 1994; unlike the Internet Society, whose members are individuals, W3C is composed of organizations. It too is concerned with the development of protocols for the Net, but at a different level. Whereas the Internet Society was established to consider the best low-level protocols of network addressing and packet routing, W3C is concerned principally with higher-level protocols for web pages and browsing. One might say that the Internet Society "owns" TCP/IP and that W3C "governs" the hypertext transfer protocol (HTTP) and hypertext markup language (HTML). The W3C was formed and has its headquarters at the Laboratory for Computer Science at MIT, at the Institut national de recherche en informatique et en automatique in Europe, and at Keio University in Japan. At the time of this writing, it was developing new standards and policies for digital signatures, metadata, rating systems, privacy, security, and electronic commerce.

In detective stories and politics, a crucial question in looking for sources of power is "Where is the money?" According to this measure of power, neither the Internet Society nor the W3C ranks particularly high on its own. Nonetheless, both organizations—through the leverage power of their memberships—are very influential. The many large corporations that are W3C members have the ability to allocate significant resources to solving technical problems—developing and promoting products and profiting from them if they succeed. From a corporate perspective, standards organizations (including the working bodies of the W3C, IETF, and the International Standards Organization) are relatively neutral forums for coordinating activities in the marketplace. Sometimes these working groups develop or promote a standard before there is demand for it in the marketplace. For example, TCP/IP, HTTP, and HTML all preceded—and in fact enabled—evolution of the Net before commercial organizations became involved. In other cases, commercial organizations develop a technology and then bring it to standards bodies with a view toward promoting it as a standard. Examples in this category include formats for pictures, audio, and video. In yet other cases, a company with sufficient market clout may try to establish its technology as a standard without regard to the time-consuming standards processes. One might argue that the digital formats of Microsoft Word, Powerpoint, and Adobe PDF became de facto (and privately controlled) standard formats for documents in this way.

Economic laws in the arena of digital goods seem to operate differently than traditional laws governing commerce in hard goods. One such difference is the so-called law of increasing returns, a phenomenon easily observed in the software industry, where manufacturing costs are so low. Arguably, this law leads to a winner-take-all marketplace in which the market leader (generally Microsoft) banishes competition. When the whole market goes in one direction, it is very difficult for competitors to gain a toehold and compete effectively. Purchasers face buying decisions on a playing field that has become tilted, and where the tilt is maintained at a level that discourages the development of alternatives. In extreme cases, the world becomes a single global digital pond with a single king frog ruling over it.

Of course, wielding such power creates its own edge and its own pushbacks. In the late 1990s, the first skirmishes about the balance of power and software monopoly were still taking shape. As always, the outcomes reflect competing values. Those on one side value an unregulated market and argue for strong "national companies" in the emerging information age. This side points to the resulting order in the marketplace and to the danger of chaos and confusion that could stem from having too many choices. On the other side are those who value competition and believe that no corporation ought to be more powerful than the U.S. Department of Justice. People on this side point to the stultifying effects of a lack of choice and innovation in the marketplace.

In the context of world history, another crucial question about the sources of power is "Who has the army?"(or as Hitler asked, "How many divisions has the pope?"). That question is one of several variants that lead to the related question, "Who has the power to collect taxes?" Traditional national concerns at borders include not only taxation but also control over materials entering and leaving a country. In the late 1990s, governmental bodies in different places and at different levels have struggled with the porosity of "sense of place" on the Net. In 1998 the city of Tacoma, Washington, was in the news for passing (and then withdrawing) a tax on Internet service providers (ISPs) that serve customers in Tacoma. It soon became clear that the "city tax" could be applied to organizations around the world. If every city on the planet assessed a 6 percent tax, the ISPs would owe far more money in taxes than they could ever earn.

So who can tax businesses on the Net? Some European countries have watched potentially taxable content providers simply move across a national border or two to avoid taxes. From the perspective of an Internet consumer, it does not matter much where a site is actually located. It is still only a click away.

The issue of content control has also loomed. Many parents have been concerned about the material their children might access on the Net. In 1996, Singapore grappled with this issue by passing regulations characterized by the press as intended to keep "its patch of cyberspace free of pornography" and to control political and religious expression. Taking such a paternalistic approach is not unusual in Singapore, a city-state that credits much of its economic well-being to careful and well-informed government regulation. But the lax controls of the Internet free-for-all resulted in a culture clash that spilled across Singapore's national border.

So who governs the Net? At one level, the question itself makes no sense. The "Net" is not a place, so it is not governed. On the other hand, many other things that are not places are regulated. Formally, international air traffic is governed by authorities in the countries where airplanes land and whose airspace they fly over. However, when it comes to air safety, the organization that sets de facto standards about the construction and inspection of airplanes is the Federal Aviation Administration (FAA) of the United States. It's not that the FAA has any official authority over its counterparts in other countries. It's just that, as a practical matter, the FAA is larger and better funded, so it makes sense to follow its lead.

Governance of the Internet—or, more generally, its leadership and shaping influence—has also tended to be largely American. It has become more international in recent years, and the international aspects are now reflected in the standards organizations and the commercial firms. As social, political, and legal issues beyond the scope and expertise of the existing bodies have arisen, specialized multinational organizations with more political connections have exercised their influence; the World Intellectual Property Organization (WIPO), for example, has become an important forum for discussions and treaties governing digital copyright. The organizations concerned with privacy and security issues on the Internet (discussed in chapter 8) are also active internationally.

The lack of centralized leadership for the Net may contribute to the perception that it is a home to anarchy, but from the practical perspective the Net is anything but anarchic. Within limits, the Net works pretty well. E-mail arrives. Web pages usually work with most browsers. Electronic commerce is starting. Other questions with more complicated answers—and requiring more work at the Internet edge (e.g., how should taxation work on the Net?)—are being answered more slowly. What's working on the Net is not anarchy and is preferable to chaos: it is a complex of institutions grappling with a complex of issues. As the title of a recent movie suggests, that may be as good as it gets.

Public Access to the Net

We tend to overlay deep social problems with technical systems. We have these deep social chasms within. One way in which they manifest themselves is with technical literacy. But technical literacy is not the cause of the social chasm. I think there is a lot of wind around providing access, but unless you address the underlying problem, I don't know that installing free computers in the local libraries will change much.

Jeff Crigler, 1997

We want to avoid creating a society of information "haves" and "have nots."... The most important step we can take to ensure universal service is to adopt policies that result in lower prices for everyone. . . . But we'll still need a regulatory safety net to make sure almost everyone can benefit.

Vice President Al Gore, National Press Club, December 21, 1993

For several years the city of Palo Alto provided two computer terminals at City Hall, where anyone could walk in and access the Net. The idea was to foster Net literacy and to reflect the social consciousness of the community. In mid–1998, however, officials removed the terminals, apparently under pressure from local businesses concerned about the impression homeless people using the terminals were making on other visitors. They sent the terminals to the public libraries. At first thought, it might seem that placing them at libraries, places we go to look up information, made sense. For homeless users, however, having the terminals there raised problems. For one thing, the libraries in Palo Alto are some distance from the business district where homeless people tend to congregate. Moreover, at the libraries access to the terminals could be subject to the same requirement that governs checking out books—possession of a library card based on a Palo Alto address. Homeless people, of course, do not have addresses. This story illustrates some of the questions surrounding free public access at the Internet edge.

But the practical aspects of public access to information and participation on the Net involve more than physical location. They can include, for example, ensuring that work stations are easy to use, that access is affordable, and that a useful set of information resources is available on-line. Nor should we characterize the access question solely in terms of the Net as an information source. Real public access would include participation in all the services and entertainment the Net offers—including e-mail and on-line discussions. In short, it would include becoming a full-fledged *netizen*—a networked citizen.

The technology and setting of the Net are evolving more rapidly than social responses to it. For example, studies of access issues I have seen cited recently are based on outdated census data and rely on Net usage patterns typical of the late 1980s and early 1990s—before the World Wide Web held sway. They often discuss models for determining payment for Net access that predate Net-based advertising and free e-mail. Assuming that the Net will continue to evolve at the same rapid rate as it has for the last decade, therefore, I propose to focus briefly on some of the underlying values, tensions, and issues I believe will become more important with the next generations of technology.

The question of how to offer the homeless access to Net terminals in Palo Alto suggests an issue that goes beyond economics, geography, and technology to a broader social question. People thinking about the homeless tend to picture them as either the "deserving poor" who need help or the disturbed, and disturbing, outcasts of society. Of course, a dualistic view of the poor antedates the Net by many centuries. The charitable urge has always competed with fears of several kinds: fear that the homeless are dangerous, that their needs are overwhelming, that we will become "like them" through association.

Today, there are social mechanisms to provide special rates to people who can barely afford telephones and energy utilities. Because television sets and radios are relatively cheap, companies and government have not seen the need to provide subsidies for the poor. But if public access to the Net was considered desirable in the late 1990s, in the next decade or two it will become even more important. If the trends cited in the preceding chapters continue, the quality of digital works available on the Net will be much higher, because digital copyrights will become real and enforceable. The amount and value of Net information are likely to skyrocket as it becomes more timely through inexpensive portable networked devices. Consumer information will become more ubiquitous as it merges in a single medium blending television, radio, newspapers, telephones, and computers. As the portable telephone and computer evolve into one portable information appliance, people will become so dependent on the Net for information that access will be essential for social participation, if not for simply getting through the day.

As the Net becomes a hub for so many digital services, the issue of subsidy will involve far more than subsidized telephone service. Digital publishers could step forward and offer discounts to the disadvantaged for access to digital works. The logic here is that providing digital copies, if not digital services, to the poor is inherently cheap, because copying a digital work is essentially free and no market share would be lost by giving a work to someone who couldn't afford to buy it anyway. Furthermore, if a formerly poor person rises above poverty, receiving such subsidies might pay off in brand loyalty to the giver. The logic behind such commercial subsidies may make them popular substitutes for public subsidies on the brave, new Net.

In short, as the Net evolves to become essential to everyday life, differing levels of service may always distinguish the wealthy from the poor. The likelihood that getting the news, banking, and communicating will all be based on portable Net devices means that the issue won't be whether public terminals are located at city hall or in the library. The deeper question will remain: How do the haves and the taxpayers see the homeless and unfortunate, and how will they choose to treat them in and out of cyberspace?

Reflections

Each time a boundary is superimposed upon reality, that boundary generates two apparently contradictory opposites. And the same thing occurs with the primary boundary. For the primary boundary severs unity consciousness itself, splitting it right down the middle and delivering it up as a subject vs. an object, as a knower vs. a known, as a seer vs. a seen, or in more earthy terms, as an organism vs. an environment.

Ken Wilber, *No Boundary*

When most of us meet someone new, our "comparing mind" becomes active to understand how we relate to that person. We may think, "we are both Christian" and therefore are part of the same flock. Or we are both Buddhist, or Hindu, or Jewish, or Sufi. The criteria need not be about religious orders. "We are both American, or Chinese." We went to the same school or work for the same company. We live in the same neighborhood. We have the same profession. We are in the same generation. We like the same music. We both have sons or daughters.

Making such distinctions and acting on them is part of the human experience. It is bound up with how we construct our sense of personal identity, how we distinguish between "self" and other. When a baby is born, psychologists say, it does not immediately distinguish between itself and its mother. Only after it develops a sense of control over its body and accumulates some experience does a baby develop the sense of a separate self.

The distinction between self and other is also bound up in our collective behavior. Nations distinguish between themselves and other nations and enforce rules at borders. Corporations distinguish between themselves and other companies, which may be customers, suppliers, or the competition. Clubs and special interest groups also do this.

The Net—like other communication and transportation technologies—unsettles and challenges our distinctions about identity. It runs right across national borders—challenging the control that countries exercise at their borders. It connects people of different social groups. By obliterating boundaries that have limited peoples' interaction in the past, the Net has the potential to destabilize some of our collective ideas about self and other.

Marshall McLuhan, the Canadian scholar of mass media, describes the *global village* as a world in which everyone is interconnected by vast networks that make communicating with people around the world as easy as chatting with a next-door neighbor. This lovely vision has inspired many people and influenced the naming of several companies and organizations. I believe, however, that the global village metaphor is fundamentally misleading.

The globe is not a village. In a village, people know each other. They share a common culture and are often closely related. Left to themselves, villages tend to maintain traditions and to change slowly. The diversity of the people, cultures, and institutions of the world make it very different from a village. The large scale of the globe means that new ideas and changes are always appearing somewhere. Global connectivity tends to spread these ideas and changes around, bringing them to places distant from their origin. From the perspective of an entrepreneur, connectivity makes it possible to do things globally that were previously only possible locally. From the perspective of a village or cultural group, connection technology brings new ways and values from other places where people have different values.

In this chapter I have explored some of the edge work we face at the boundaries in our world: the Net in first and third world countries; the perception of cultural mixing and cultural pollution; the broad effect of the Net on democratization; the cluster of organizations trying to govern the Net and shape its future development; the separation and mixing of haves and have-nots on the Net.

People have different reasons for using the Net. Many companies are trying to expand their businesses by going on line, while individuals often connect their computers to the Net so that they can explore the world. Neither necessarily recognizes consciously any erosion of boundaries. Yet, by using the Net and increasing their connectedness with other people, they are altering the boundaries that they perceive.

The Net is not going to transform the world immediately into a unified place. We have many differences in our cultures and values. Travel is broadening, but culture shock is visceral. Embracing too much at once is overwhelming. Even as we connect ourselves together, we do other things to preserve our diversity. But our curiosity and urge for growth and experience is bringing about new mixtures. We are part of a world that is trying to understand itself. The Net is facilitating that exploration.

10

Indistinguishable from Magic: The Real, the Magic, and the Virtual

One might say that our culture is "possessed" by the immature shadow-magician. When human beings use their magician potentials in the service of healing and community, the . . . energies of the immature magician—the trickster—are transformed into a mature, shamanic form that heals both self and the larger community.

Robert L. Moore and Douglas Gillette, *King, Warrior, Magician, Lover*

Many of the smart technologies that surround us—computers inside car engines, appliances, houses, offices, traffic lights, and many other objects—resemble the visions of magic glimpsed in archetypal myths and folk tales. Handheld remote controls are our magic wands; we push the button and the environment responds automatically—opening the garage door or changing the music on the CD player. Smart automobiles and programmed trains perform as magic carpets; without waiting for directions, they whisk us away to our destination. Like the enchanted castles of fairy tales, smart houses and buildings respond to our every command, turning up the lights or lighting a fire in the fireplace. Even the small rings or electronic badges we wear for security or identification are enchanted, the technological equivalent of magic rings. They tell sensors in the environment that we are entitled to the powers, access, and privileges of an select group. Like the magical cloaks and suits of ancient folk tales, wearable computers monitor the environment and bring us information just when we need it. Even the lowly television set, when connected to the Net, becomes a magic mirror and a crystal ball, displaying objects and people far away and letting us interact with them.

Through its powers of computation and communication, the Net increasingly gives the machines and devices of our physical environment the capacity for intelligent action. Networked, they can act with coordinated purpose and respond to remote or local control. More and more, the computerized world harkens back to the magic lands of our childhood memories, the Emerald City of the *Wizard of Oz* or the enchanted forests of Grimm.

Like those magical locales, smart technology raises issues about power and control. Technology can offer us almost-inconceivable levels of control, or it can threaten to control us. With so many devices in our environment being computerized and networked, we find the world to be simultaneously real and virtual, magical and mundane. Such a world can be friendly and serving, or sinister and observing. Video eyes in the environment can watch over and take care of us, or they can spy on us and record our words and actions. The challenge of magic in every age is not just to develop and use new technologies but to understand how shifts in power and control can best serve ourselves, our communities, and our global environment.

Once Upon a Time

The relationship between high technology, magic, and fairy tales has been noticed by many writers. Arthur C. Clarke, the futurist, scientist, and science fiction writer, wrote that "any sufficiently advanced technology is indistinguishable from magic." Clarke, the author of over seventy books of fiction and science, was the first to propose the geosynchronous earth orbit on which all modern communication satellites are based. He is perhaps best known for his screenplay (written with film director Stanley Kubrick) for the movie *2001: A Space Odyssey,* which is based on the germ of an idea in his short story "The Sentinel."

Even scientists and technologists usually considered hard-nosed and practical recognize the kinship of science and magic. Clarke's notion of advanced technology was similar in spirit to that of Allen Newell, the psychologist and computer scientist. Newell was widely respected as a hard-evidence, no-nonsense experimentalist and theorist whose contributions range from innovations in computer design to models of human problem solving based on information processing. He thought a lot about the practice of science and even wrote rules for his students to help them manage their careers. He especially advised them not to speak in public about wildly speculative theories, except on rare ceremonial occasions.

One day in 1976, on just such an occasion, Newell gave an uncharacteristic address. The event was the inauguration party for the U. A. and Helen Whitaker Professorships at Carnegie Mellon University, at which Newell was awarded an endowed chair in computer science. (He held it until his death in 1992.) In his acceptance speech, Newell pointed to the close relationship between fairy tales and technology: "The aim of technology, when properly applied, is to build a land of Faerie. . . . the computer . . . [is] the technology of enchantment" (Newell 1992: 46).

Essential to the advanced technology Newell wanted to build were two ingredients: a capability for intelligent action and physical miniaturization. In other words, the technology needs to know what to do. Technology does not seem very magical if it does nothing very useful or very well. Secondly, technology needs to be small enough to disappear. When we can see the spinning gears and shafts of the mechanism behind it, no machine appears magical. Aware of Moore's Law predicting continued advances in computer technology, Newell was confident that computers would become ever smaller, cheaper, faster, and more reliable.

In his speech Newell cited as an example of "enchanted" technology antilock brakes, which stop cars safely on wet pavement. The brakes are "smart" enough to sense slippery roads and small enough to disappear into the car. At a more speculative level, he imagined talking streetlights that would assist people standing under them, perhaps asking "May I help you?" Streetlights programmed with the map of the city could give directions or summon help for people who are lost or in trouble.

Newell recognized the dark side of his fairy tale analogy too, pointing to three folk tales from which technology writers often draw troubling conclusions.

> The Sorcerer's Apprentice learns only enough magic to start the broom of technology hauling water from the River Rhine to the cistern but cannot stop it.
>
> For the Jinni in the bottle, the story is never permitted to go to the conclusion in the Arabian Nights, with the Jinni snookered back into the bottle; it is always stopped with the Jinni hanging in the air and the question along with it: Can we ever put the Jinni back, or will there only be ink all over the sky 'til the stars go out?
>
> In the many stories of the three magic wishes, promises of infinite riches are just for the asking, but the wishes are always spent, first on foolishness, second on disaster, and third on bare recovery. (Newell 1992: 46)

At first glance, the cautionary tales cited by Newell predict disturbing fates for those who rely on enchantment, but, like all enduring myths, they lend themselves to deeper understanding. Such stories have been studied by Joseph Campbell (1988), Bruno Bettelheim (1977), Clarissa Pinkola Estés (1992), Marie-Louise von Franz (1972), and many others. If high technology is, as Clarke and Newell suggest, indistinguishable from magic, it is arguably worth mining these old stories about misused magic for advice about our own relationship to technology.

In the "Sorcerer's Apprentice," the technology runs amok. However, when the wise and powerful sorcerer discovers the mess, he does not break the brooms or blame the technology. Instead, he admonishes his apprentice and requires him to mend his ways. His original assignment to his assistant was "Chop wood, carry water." This instruction—eerily identical to one from ancient Zen masters—is about creating mindfulness in everyday activities. By trying to avoid the work of building consciousness, the assistant created a disaster. Among other things, the apprentice needs to learn awareness, patience, and responsibility.

Bettelheim suggests two possible interpretations of the folk tale of the Jinni and the Bottle. It may be about "bottled up" feelings that can get out of control, or it may point to the need for children (ordinary people) to use their wits when dealing with adults (giants). Newell complained that in many discussions of technology the original story is diminished by leaving out the conclusion in which the child tricks the Jinni back into the bottle. ("It's hard to believe that a big and powerful jinni like you could ever fit in so small a bottle!") The truncated story thus becomes a cautionary tale, a metaphor for a bad invention—such as the computer or the atomic bomb—that, once created, cannot be uninvented. The apparent moral is that we need to take responsibility for our creations. If we don't, or can't, then some things are better not invented at all.

The various tales of the Jinni in the bottle are often taken to mean that automation can be dangerous. When our wishes are fulfilled by machines that mindlessly follow our bidding, we do not observe the effects and do not appreciate what we have. We may wish for things casually without realizing that the machines, our unthinking servants, are polluting our air and water, digging up the earth, or cutting down our forests—just to fill our lives with gadgets we don't need. In an increasingly automated world in which machines do much of the hard work, the worrisome result may be wastefulness and mistreatment of the environment.

In the following sections, we consider several tales about technology from a contemporary point of view. All the stories are variations of two familiar, polarized themes: building a technological utopia (Newell's "Land of Faerie") and warnings of a rampaging technology. As the capabilities of our new technologies come to approximate more and more closely the magic of fairy tales, we need to consider how we can apply the lessons of the magical literature about power and control to the world we live in.

Beepers Go Berserk!

Fan-out, one of the powerful properties of electronic communication technologies, also creates opportunities for widespread disaster. When things go wrong they can go very wrong very quickly over a large area. On January 10, 1997, a *Wall Street Journal (WSJ)* front page story described the effects of an apparent technical malfunction of the SkyTel paging network. In the resulting bout of "beeper madness," a deluge of call-me-back messages went out to more than a hundred thousand pager customers.

What went wrong? It started when a SkyTel customer was trying to reactivate service with her new seven-digit identification number. Somehow, the SkyTel employee had assigned her a number that was linked to a SkyTel secret code used by the company to beam out news headlines simultaneously to a hundred thousand of its 1.2 million customers. The SkyTel computer resisted the number assignment, but staff members successfully overcame the computer's "recalcitrance" and assigned the number anyway. When the customer tested it, the wireless chain reaction began: a wave of messages went out to a hundred thousand customers.

SkyTel beepers display a seven-digit number, which can be read as either a local telephone number or a SkyTel identification number. Hundreds of customers all over the country interpreted the number appearing on their beeper as a local phone number, resulting in thousands of calls to puzzled folks in different area codes. But the big chain reaction was yet to come.

When about three dozen SkyTel customers—the article called them the "diligent three dozen"—decided that the seven-digit number they had received was not a local telephone number but a SkyTel identification number, they sent their own phone numbers to the magic number. SkyTel then beamed their phone numbers out to the same hundred thousand customers. Thousands of SkyTel subscribers then telephoned those three dozen diligent people, overloading the circuits and filling voice-mail boxes. The newspaper article chronicles the woes of several of them, such as an AT&T Corporation manager who received 625 calls from perplexed strangers.

What, ultimately, do we make of the mess? John Kirkpatrick and Jared Sandberg, authors of the *WSJ* article, poke fun at the beeper users, whom they characterize sardonically as the "hopelessly in-touch." But were the beeper users, especially those who recognized the number as a SkyTel PIN and called it back, really at fault? Did they, as the journalists suggested, have "only themselves to blame" for being so "devoutly wireless"?

A couple of months after the news article appeared, I noticed the following advertisement in a later issue of the same paper.

A FORMAL APOLOGY
Since inventing cellular
and after introducing
digital wireless,
wireless office
systems and cordless phones,
it seems that
anyone can get ahold
of you no matter
where you are.
Sorry.

Lucent Technologies (Bell Labs Innovations), *WSJ*, March 14, 1997

This tongue-in-cheek advertisement implicitly raises the same question of blame. It modestly suggests that the technology is at fault, not those who are hopelessly in-touch. That same day I needed to return a telephone call to a publicist at a large corporation. Her voice mail gave me a paging number, which I called. When she called back a few minutes later, I asked her whether—with her cell phone, pager, voice mail, and e-mail—she was hopelessly in-touch.

Her reply, while not as witty as the alliterative prose about the digitally diligent, brought me a deeper understanding. She is a working mom with kids. To meet her family responsibilities she needs to be mobile, and to meet her work responsibilities she needs to be quickly available to media people and other contacts. The communications technologies give her accessibility combined with freedom of movement. From her perspective, beepers and cell phones are empowering technologies. They help her to balance her responsibilities and use of time, making life better than it would be without them.

It is too simple, therefore, to complain that we are excessively in touch, or that technology is sure to run amok, or that there will always be human error. We want the new technologies because they enable us to do things that fit in with our lives. Still, the beeper incident and our recognition that all applications of technology entail some risks bring other questions to the surface. Are we becoming too vulnerable to technological failure? Given that communications technologies with fan-out amplify the effects of our mistakes, are we aware enough of the possible danger and putting enough thought into limiting damage if our gadgets do run amok?

In May 1998, another potentially catastrophic failure of beeper systems provided an opportunity to consider this question. That day, a computer in PanAmSat's Galaxy IV satellite failed and the satellite rotated out of position, blocking communication to thousands of antennae at ground stations. Although the satellite was used for many kinds of communication services, one of the most noticeable effects of its failure was the silence that fell as beepers all across the United States shut down. Because satellite time is expensive and paging uses very little capacity, eight of the ten largest beeper companies had all rented space on Galaxy IV.

The overall effect on service depended crucially on whether the providing company had a backup channel. Although the CBS, UPN, and WB networks also use Galaxy IV to beam programs to their affiliates, they were mostly able to switch automatically to alternate satellites when Galaxy IV went dark. The paging networks, however—being based on a single-point-of-failure, low-cost approach—remained silent for several days.

Beepers and paging systems receive signals not directly from the satellite but from local radio transmitters, which rely on the satellite to shuttle messages between cities. Transmitters communicate with the geosynchronous satellite by dish antennae pointed at the right part of the sky. When Galaxy IV went dark, paging companies had to buy space on another satellite and—a more time-consuming task—manually reposition thousands of antennae in different cities to aim at a different satellite. Over the next few days, service was restored one city at a time.

Many of the news stories about beeper failure focused on situations in which people had depended on the reliability of the service to organize their lives. In hospitals, for example, beepers are used routinely to page doctors in crisis situations. When Galaxy IV went dark, doctors had to rely on wired phones and cell phones. As a result of this incident, beeper companies may offer multiple levels of service: an inexpensive service for the mass market and pricier but more reliable services for high-value and emergency applications.

Wearable Computers

Slip on the ring and your index finger becomes a scanner; slip on the WSS 1000 and your forearm becomes a workstation.
On-line advertisement for Symbol Systems products.

In certain folk tales magical clothing conveys special powers to the hero—cloaks make them invulnerable to fire or sword or give them the ability to fly or to walk invisibly through the world. Wearable computers, similarly, are being designed to give wearers heightened powers to sense things in the environment, to receive information, or to control their environments through remote connections with distant computers.

In the early 1990s, research programs in wearable computers were started at the Massachusetts Institute of Technology, Carnegie Mellon University, Columbia University, Georgia Tech, Xerox Palo Alto Research Center, and several other places. Start-up companies and defense contractors also entered the picture. By the late 1990s, the technology of wearable computers was rapidly developing miniaturized applications like the special eyeglasses that allow the wearer to see the real world or to focus on information displayed by the glasses. A Boston start-up company called MicroOptical has designed these ordinary-appearing eyeglasses, whose frames conceal a display and other electronics. When the display is turned on, the wearer sees a virtual video or a computer screen seemingly floating about three feet away. The image, which is created by a small liquid crystal display, is sent through the lenses and then reflected by microscopic embedded mirrors.

A bulkier version of the eyeglasses is a head-mounted display built by Virtual Vision. It also includes microphones, earphones, and speech-recognition software, providing hands-free interactive applications and allowing the user to both access information and give spoken commands. In addition, the helmet includes an embedded camera, so that it can support either video recording or eye-tracking, that is, directional control guided by the wearer's eye movements. Another variation of this technology, a one-inch eyepiece worn on a headband, was developed by Rockwell.

These devices can simultaneously show the user the real world and a superimposed data display. Augmented-reality applications of such devices were anticipated by science fiction film writers several years ago. The *Star Trek* series, for example, featured an augmented eye loop for medical procedures; the loops superimpose data about the patient's condition over the doctor or nurse's view of the patient.

One issue encountered in designing head-mounted displays is the difficulty of providing a "mouse" equivalent, that is, a means for wearers to point to particular objects they want to examine in the virtual or physical displays. One approach is to build into the glasses eye-trackers, which move the cursor to wherever the wearer looks. Other alternatives include belt-mounted trackballs, wrist-mounted thumbwheels, and finger rings whose position can be discerned by the magnetic sensors. Keyboards have also been woven into clothing and mounted on an arm bracelet. Symbol Technologies makes rings that can scan bar codes and radio-linked computers and keyboards that can be worn on the forearm. A variety of other sensors—including some that monitor users' galvanic skin or muscular responses or their respirations—are being employed experimentally. Sensors that scan the external environment include, among other devices, video and infrared cameras.

The similarity of wearable computers to the magic rings and cloaks of ancient myths and, even, comic books, has not been lost on vendors. One of the more colorful wearable computers is based on the iButton, or information button, made by Dallas Semiconductor, best known as a maker of microchips. One version of the ring containing applications written in the Java language has been promoted by Scott McNealy of Sun Microsystems. In an ad, McNealy is shown flying about with a red cape, his fist extended like Superman's to display a "magic" ring that radiates sparkling beams. Next to the picture—only partly tongue in cheek—the promotion announces "Javaman: The Adventures of Scott McNealy. Today's Episode: His Fight to Save the World Wide Web from the Evil Empire." Another proposed ring device that continues the comic book theme is the decoder ring—a cryptographic application based on a microprocessor, math accelerator, secure memory, and tamperproof clock, all built into an iButton. Another application of the iButton is an electronic wallet capable of carrying electronic cash and equipped with all the usual smartcard functions for electronic commerce. Lest the association with power and control be missed, the iButton comes with an "I" (not an "eye") on its face, emphasizing the slogan "Information is power."

Where then is the computer behind the magical ring, or wrist keyboard, sensor, or display glasses? One possibility offered by ViA Systems is a bendable computer built into a belt. Spotting the product at an electronics show and recognizing the comic book imagery, Bill Gates reputedly said, "Batman would feel right at home wearing one of these."

A major engineering challenge for the experimental wearable devices is how they should communicate with each other when someone is wearing several devices. Using wires or cables to connect them adds extra steps when the wearer is donning them and could result in tangled wires. One alternative is to build conductive fibers into the wearer's garment so that it can be used as a "data bus." Another proposed approach—called the personal area network (PAN) or body area net (BAN)—uses the human body as an electronic conduit between wearable or portable devices.

Going beyond even computer-enhanced clothing and using the body as an information conduit, some researchers are engineering devices to fit into our bodies. In effect, this follows a progression inwards—engineering our environments, engineering our clothing, and engineering our bodies. As Neil Gershenfeld, a future-thinking technologist at MIT, has quipped, "From this angle, 'Intel Inside' takes on a whole new meaning."

Funding for much of the development of wearable computers comes from the Defense Advance Research Projects Agency (DARPA). In the late 1990s, the Warfighter visualization research at DARPA focused on three main areas: visually coupled systems, geospatial registration technologies, and enhanced human–interface technologies. *Visually coupled systems* obtain data from the environment through sensors and give wearers a view of that information to enhance what they see with their own senses. *Geospatial registration technologies* allow a computer system to determine the exact location of an environmental object or to superimpose a representation of known objects onto a field of view. *Enhanced human–interface technologies* are designed to present information to users in a way that is carefully adapted to what is known about human cognitive abilities. In addition to their military applications, all of these research areas have potential uses for living and working in augmented realities.

The metaphors and names we use for technologies can be a clue to how we imagine using them. The name Gladiator™ from Symbol Technologies, for example, suggests a warfare application, while the name Trekker™, from Rockwell, indicates that the system is useful for a long journey by foot. The Xybernaut™, made by a company of the same name, implies a relationship to astronauts or aquanauts, who wear suits in inhospitable environments. (The *Xyber* prefix suggests that it is used in cyberspace.)

The military applications of wearable computers also bring us face to face with issues of power and control. These issues are often at the core of stories about magical rings, weapons, or clothing. In J.R.R. Tolkien's *Lord of the Rings* (1966), for example, the top ring of power amplifies the powers of the wearer, making someone of even modest personal power invisible. However, the ring has a purpose of its own and slowly bends the wearer to that purpose. In Hans Christian Andersen's *The Red Shoes,* the ballet shoes give the wearer the ability to dance fabulously. But the shoes too have their own mission, and the wearer finds she cannot stop dancing. In Dr. Seuss's *The 500 Hats of Bartholomew Cubbins* (1938), the wearer is supposed to doff his splendid hat when the king goes by. However, every time he takes it off, yet another hat appears on his head. In the classic mideastern tale, *Aladdin and His Magic Lamp,* Aladdin finds a lamp inhabited by a powerful genie. The genie obeys the commands of whoever possesses the lamp, but when Aladdin loses it he must get by on his own wits.

These stories raise several issues and recurring themes that are part of the edge for wearable computers. One theme is that the magical clothing or appliance may have a will and purpose of its own. In *Lord of the Rings,* the unwarned wearer, blind to the purpose and influence of the ring, unconsciously exercises its power until its enslaving purpose becomes his purpose. A similar difficulty with a less subtle effect is illustrated in *The Red Shoes,* in which the question becomes whether the dancer can ever take off the shoes. In a modern variation, the Borg of the *Star Trek* series are so deeply interconnected by embedded and wearable devices that their individuality is lost. Indeed, the unitary nature of the collective is so complete that the term *Borg* does not even end with an *s*. In all these stories, the magical powers affect the very selves of the wearers. The greater the power, the greater the influence and risk.

A second recurring theme is that the wearer can become too dependent on the device. In *Lord of the Rings,* earlier wearers of the ring, such as Gollum, still crave its power and scheme to get it back. In *Aladdin,* when the protagonist loses the lamp, he also seeks its return. The stories thus show us protagonists in contrasting situations—living with magically augmented powers and depending on their own abilities. The more powerful the device and the longer wearers depend on it, the weaker they become on their own. A warning here is that a user of magical devices should not become overconfident or too dependent upon them. In the real world of failure-prone technologies, the risk is not only that the device will be lost but also that it will malfunction—either failing completely or, perhaps more confusingly, providing false information.

Even when devices are working according to design, users need to divide their attention between what their own wits and senses tell them and what the device is displaying. A contemporary device, the cellular phone, has become indispensable for certain lines of mobile work—such as real estate sales. It has also seen as a major cause of automobile accidents that occur when drivers divide their attention between traffic and a remote conversation.

Moreover, the same technologies that give us greater mobile access to data about the environment can give others greater access to data about us, enabling them to monitor our personal lives. In dangerous situations, such monitoring could provide an additional measure of security—summoning help or providing timely advice. In other situations, such as when a supervisor covertly observes his or her subordinates or reads their e-mail, the devices are potentially intrusive.

As is the case with beepers, we can draw no universal conclusion about the values and risks of wearable computers. There are appropriate and inappropriate applications. For some kinds of work they will undoubtedly have great benefits. Unlike the computers we leave behind at our desks, wearables are intimate devices; they could be with us all the time, recording information, providing assistance, and keeping us in touch with otherwise invisible aspects of the world. But whenever we pay attention to information from the device, we necessarily shift the focus of our minds and senses away from our immediate environment. In any given context, we need to ask whether the information displayed by the device is worth more to us than what we perceive through natural means. We can also ask whether the potential for increased surveillance is life enhancing. The Internet edge for the continuously wired existence involves weighing the competing potential values and risks of such connectedness and technological intimacy.

ToastScript and the Networked Home

Imagine coming home at the end of a hard day and having the whirlpool tub already filled with water at just the perfect temperature to bubble away all your stress and tension. Imagine controlling the lighting, the temperature and the "mood music" throughout your home with no more effort than it takes to change channels on your big-screen TV. Then, imagine yourself as the new owner of this fully computerized luxury home.

Advertisement for a smart house in New Mexico

In the late 1960s, the first elements of home automation technology grew out of hobbyist experiments. In the mid–1980s, several industrial groups began to pursue ways to integrate automated household devices through centralized control systems. The more data smart devices have about the context—including the people, their needs, and the situation in the house—the more things smart homes can, potentially, do. Except for customizable parameters such as preferred water and air temperatures, most centralized automation systems currently available for residences have access to only limited information about the outside world—usually the time of day, temperature, and power company billing data. In the late–1990s, however, technology designers began moving toward providing enhanced information and control possibilities by connecting home devices to the Internet.

The much greater opportunities and risks involved in such connectedness have inspired modern jokes and cautionary fairy tales that harken back to stories of enchanted castles and haunted houses.

Home Automation Today

When you plug a smart object into the house-net, its chip declares its identity ("I am a toaster"), status ("I am turned on"), and need ("Give me 10 watts of 110"). A child's fork or broken cord won't get power.
Kevin Kelly, *Out of Control*

X–10, a protocol for controlling home devices, was originally developed by Pico Engineering in the late 1970s. Products compatible with the protocol are now offered by several companies including X–10 Powerhouse, Radio Shack Plug 'n Power, Stanley Lightmaker Home Control, Heath/Zenith, and Leviton's Decora Electronic Controls.

In its simplest form, X–10 can control appliances by turning their power on or off. A lamp or other appliance plugs into an X–10 module, which plugs into a wall outlet. Each module assigned to an appliance has a unique identifier, such as A–2, B–3, C–4, etc. On/off signals are sent through existing house wiring. A signal such as "A–2 ON" would be ignored by all modules except the lamp module with the A–2 identifier. Other commands in the protocol suitable for dimmer switches would include BRIGHTEN and DIM.

Several control options are available on an X–10 system: tabletop controllers that plug into any AC receptacle can send X–10 signals throughout the house; wireless controllers send radio signals to a plug-in base transceiver, which then sends appropriate X–10 signals through the electrical wiring. Sensors that can also send signals—such as motion detectors, clocks, temperature sensors, light sensors, moisture sensors, and sound sensors—are also available. Computer- and telephone-interface devices are other control options.

In 1984, the Consumer Electronic Bus (CEBus) Committee of the Electronic Industries Association (EIA) formulated a communications protocol for networking residential electronics. The CEBus Industry Council (CIC) proposed a common application language, called Plug 'n Play (PnP), which could be used with any signal-transmission mechanism. PnP is intended to facilitate communications to a wide range of home devices, including television sets, lamps, electric door locks, dishwashers, hi-fi equipment, drapery cords, burglar alarms, and so on. Members of CIC, an international support organization for companies creating home automation networks, include Ademco, AMP Incorporated; AT&T, Compaq Computer, Hewlett-Packard, Honeywell, IBM, Intel, Leviton Manufacturing, Lucent Technologies, Microsoft, Molex Inc., Panasonic, and Thomson Consumer Electronics. Unlike the X–10 system, the CEBus protocol could apply to a broad range of transmission media including power lines, twisted-pair wires, coaxial cables, infrared, and radio.

The market in home automation stalled during the 1990s, in part because of confusion surrounding the competing standards proposed for interoperable products. For example, Smart House, Inc. offers a proprietary protocol for the new construction market intended to support control of television, audio, telephone, power distribution, natural gas distribution, and home automation. It places convenience centers in various parts of the house with outlets for power, data, telephone, and coaxial cable. A competing approach to home automation is LonWorks, which was developed by a start-up company, Echelon Corporation. LonWorks is designed to scale up or down for everything from family homes to large commercial, industrial, transportation, and medical installations; it can accommodate tens of thousands of devices. It has its own control protocol, LonTalk, which can be used with many different transmission media. Chip-based devices supporting LonTalk are made by Motorola and Toshiba.

Working at a higher level than the individual residence applications, several electronic companies and power companies have experimented with using the power-distribution network to carry voice and data in addition to electricity. In the United States, the Novell Corporation and Utilicorp United Inc. have announced development of Novell Embedded Systems Technology (NEST), technologies that could transfer data over power lines at speeds up to two megabits per second. The value of such a smart energy system for power companies is that it would gather data to be used to manage power loads and to shift nonessential uses of power to hours when power is cheaper. Commercial trials of this technology began in the late 1990s.

Home Automation Meets the Net

The premise of virtual reality is to put you inside a computer world. I want to do the opposite. I want to put the computer world around you on the outside.
Mark Weiser, "The Computer for the Twenty-first Century"

In the late 1990s, interest in linking the Net to home automation and other devices for controlling the environment led to the development of new versions of Internet protocols for linking devices. The slogan of the World Wide Web conference in 1997—"Everyone • Everything • Connected"—refers to devices in the home as well as to the people of the planet.

Connecting home devices to the Net opens the door to applications that do more than turn lights on and off, support security systems, and prepare the bath water. The key idea is to expand the external information available to home automation systems beyond parameters like power consumption, time of day, and temperature. Such Net-connected applications could include a much wider range of information services: for example, sensors in microwaves and other home appliances that could read product codes on food labels and look up cooking instructions on the Net; refrigerators that could conduct inventories automatically, consult vacation schedules, and automatically suggest ordering new food and discarding food whose shelf life has expired.

In the mid–1990s, early versions of office systems and information appliances that communicated through the Net were built as research prototypes. An example of such a project was the Ubiquitous Computing Project at Xerox PARC, headed by Mark Weiser. Weiser envisioned an office containing hundreds of small devices, all linked. There would be interactive boards of different sizes—ranging from post-it-sized notes to blackboard-sized surfaces. Even books could have tiny chips embedded in them. Individuals would carry small badges linked through infrared beams to let the Net keep track of where they were. The project built operational prototypes of many of these devices—including badges and communicating handheld personal digital assistants (PDAs)—and installed infrared transceivers over a substantial part of the research center. The Ubiquitous Computing Project helped members of the Internet community understand the need to extend Internet protocols to support lots of small, networked devices.

Over a period of several years during the 1990s the number of network systems connected to the Internet doubled every twelve months. The version of the network protocol used in the early 1990s is designated IPv4 (Internet Protocol version 4). The new version is called IPv6 or IPng (for next generation); in September of 1995 it became the proposed standard. By the late 1990s, there were implementations for over thirty kinds of systems. As this is written, use of IPv6 is limited mainly to research organizations. Deployment to the majority of sites on the Net is expected to take place over several years.

IPv6 is designed to run well on high-performance networks as well as low-bandwidth networks such as might be available with wireless or power networks for home devices. It is also intended to accommodate expected growth in the number of Net-connected personal computers, portable or nomadic computing devices such as PDRs, networked entertainment vendors, and devices such as lighting equipment, heating and cooling equipment, and home appliances. Members of the Internet Engineering Task Force (IETF), which designed IPv6, hope to offer a ubiquitous standard to meet this broad range of emerging markets and to create a worldwide, interoperable information infrastructure with open protocols.

A serious limitation of the present, IPv4, protocol is its reliance on thirty-two-bit addresses. This address-size limit seemed very generous when it was proposed and is probably large enough to carry the Net into the first few years of the twenty-first century. IPv6 was designed to expand addresses to 128 bits, a number large enough to support 665,570,793,348,866,943,898,599 devices per square meter of surface area on the planet.

IPv6 also includes many other extensions and improvements over earlier Internet protocols. Two of the improvements deal with security. One is an authentication header intended to prevent tampering with the contents of communications and improve information about the sources of data. A second extension, the encapsulating security header, provides the same functions but in a more complex manner designed to ensure that information is kept confidential.

Fairy Tales about Home Automation

I want to go with Kitchen 98 unless I am given an incredible bargain on a refrigerator equipped with Kitchen 95. It looks quite similar and the cost differential for that feature is minimal. Upgrade software is I think inferior to getting it right the first time. Besides it means redoing the upgrade any time your system crashes.

Minor word substitution on a communication to the author from Eric Stefik about buying a computer, 1998

The emergence of devices that are both intelligent and interconnected promises a new world of home convenience. Back issues of *Popular Science* are full of ideas about automatic control of home appliances, but a perennial problem with many of these contrivances is the method used to tell the devices what to do. I once read an e-mail message to a news group on home automation that described a late-night party held at the home of an automation enthusiast. At 10:30 the lights went out automatically, although the party wasn't over. To restore the power, the owner had to go off for what seemed an inordinately long time. Of course, the house could have been preprogrammed to support a party mode as well as early-to-bed mode. Alternatively, there could have been motion sensors in the room to put the lights back on when something moved, although that wouldn't work very well if the motion was caused by a breeze blowing the draperies or the cat going to its food bowl in the middle of the night. Even if homeowners programmed the lights to go on only when someone gets up for a drink of water, they wouldn't want them at full brightness. This example illustrates that, to be intelligent, a home-appliance network needs a great deal of context-specific information, and that getting information into the network requires careful programming. Setting the controls just right is not easy because there are always special cases and exceptions to provide for. As the old proverb reminds us, a little knowledge can be a dangerous thing.

People's experiences with the programming of complex systems have inspired many cautionary tales. In 1993, anticipating the advent of home automation, Michael Schrage published in the *Los Angeles Times* a spoof on home automation gone bad. The story has become a classic and can be found on many web sites. It starts out as the owner moves into his new automated home and begins enthusiastically tuning it to his specifications. The first bugs surface after a month as unanticipated interactions among the appliances lead to a situation that is difficult to diagnose. There is much finger pointing between the different vendors and utilities as new glitches continue to surface. The climax comes while the homeowner is waiting for release of the new, improved version of the system software.

> This is a nightmare. There's a virus in the house. My personal computer caught it while browsing on the public access network. I come home and the living room is a sauna, the bedroom windows are covered with ice, the refrigerator has defrosted, the washing machine has flooded the basement, the garage door is cycling up and down, and the TV is stuck on the home shopping channel. Throughout the house, lights flicker like stroboscopes until they explode from the strain. Broken glass is everywhere. Of course, the security sensors detect nothing. I look at a message slowly throbbing on my personal computer screen: "Welcome to HomeWrecker!!! Now the Fun Begins . . . (Be it ever so humble, there's no virus like HomeWrecker . . .)" I get out of the house. Fast. (Schrage 1993)

The insurance policy, of course, doesn't cover damages because it doesn't mention computer viruses, and the warrantee for the home automation system has exclusions for systems connected to uncertified services on the Net.

Such cautionary tales are epigrammatic discussions of values in conflict at a technological edge. As Schrage's story begins, the automated house offers the owner the alluring convenience of turning over control of ordinary household functions. It seems like an enchanted castle but turns out to be a haunted house that brings horror and loss of control. In this regard, it has parallels with the red shoes and the magic ring that have agendas of their own. The deficiencies in state-of-the-art computing are bad enough when confined to our computers, where they merely interfere with work. When they live with us in our homes, up close and personal, their potential to terrorize us is greatly amplified.

Schrage's story chronicles several familiar problems of automated systems: the incompatibilities of software from different manufacturers, the lack of testing in all the special conditions that could arise, and the susceptibility to viruses. Similar concerns are reflected in on-line jokes about the incompatibility of various operating systems and toasters. We dream of just plugging things together and watching them work together smoothly. In the current state of the art, realization of that dream is still remote.

Malevolent computer viruses and incompatible software are just two of the risks encountered. Even home automation software that is apparently bug free may best represent the interests of people other than its user. In the scenario of one playful networker, a man turns over a certain amount of control of his home to outside businesses in exchange for credits and discounts. Other network wags, punning on the name of the PostScript™ language used in the printing industry, have devised ToastScript. The idea is that home-automated toasters, with their array of heating elements, can heat or thermally print tiny dot patterns on pieces of toast. In this way, messages—presumably communicated through the smart plug attached to the toaster—could be printed on toast. Imagine a slice of toast popping up in the morning with various printed and audio messages: "This toast brought to you by the Milk Council. Wouldn't a glass of milk or hot cocoa taste great right now?" If the smart toaster communicated through its power cord with the smart refrigerator, it might print a list of the household's supply of breakfast foods.

It may not be too far-fetched to imagine that convenience, and the opportunity to earn commercial credits, might persuade us to let our appliances communicate with servers on the Net. The refrigerator could keep an inventory of goods, their consumption, and their shelf lives and use it to order replacement items automatically. Ice cream companies could adjust the temperature of refrigerator compartments to optimize the storage and consumption of ice cream. Loyalty programs for Brand X ice cream could offer us electricity credits if we let Brand X systems control the refrigerator settings. A kitchen drinks dispenser could be programmed to recognize whose cup is under the nozzle and fill it with that person's beverage of choice—keyed to the time of day and, possibly, favoring the products of Brand X. In the late 1990s, a project called Kitchen Sync at the MIT Media Lab designed prototypes of similar products; it envisions providing the cook with verbal and video aids and coordinating the appliances as they measure ingredients and cook the meal.

Automatic systems present plenty of possible pitfalls. To avoid days of answering silly questions or confirming repetitive actions by machines, we might be tempted to let the machinery carry out certain actions automatically without asking for human authorization. Imagine a home kitchen system that orders food while a family is on vacation, watches the food spoil, and then discards it because nobody told the house the residents would be gone for a week. With quiet inefficiency, a not-so-smart technology could wastefully carry out its work unnoticed, cleaning up the evidence of the wastefulness and paying the excessive bills automatically.

There are also opportunities for warfare between competitive suppliers of products and information. Brand X might try to control cooking or cooling cycles in a way that enhances its own products but causes products from Brand Y to be overcooked or to spoil. Imagine a kitchen cooking a dinner that malfunctions because the dinner is from Brand Y and the microwave is controlled by Brand X.

A system that accepts instructions from many sources thus has the potential to be very difficult to understand and debug. The smart utilities in an automated apartment, for example, probably would be programmed by multiple instructions from the landlord, the utility companies, and the renter's association. The landlord may install instructions that preserve the value of his or her investment in the dishwasher, while utility companies may program appliances with instructions to regulate power usage. The principal concern of the renter's association may be to minimize noise for neighboring tenants. All these instructions can have subtle interactions with compounding and contradictory effects. What would happen if the utility company instructions to run the dishwasher at low power in the middle of the night (to reduce the peak electrical load) are incompatible with the renter's association instructions to use it when people are not sleeping and with the landlord's programming to run it at higher power periodically in order to prevent build-up of residue? The usual kinds of interactions would probably get worked out routinely, but unpredictable and unusual cases—like the tenant trying to run a load at an odd hour to accommodate a quirk in a personal schedule—could lead to frustration when the dishwasher won't run at all. When an appliance fails to work as the tenant expects it to, it may be very difficult to identify the source of the problem and impractical to change the programming.

Another dark side to allowing home appliances to monitor and report on activities in the automated house is the potential for unwanted surveillance. This issue could arise from the increase in telecommuting and the concern some employers may have that employees do not work as hard when they are not being observed. For example, access to employees' usage logs could indicate that they are ordering movies to play in their home office at a time when they should be working. A medical insurance company, similarly, could check up on whether a client was eating foods consistent with a medically prescribed diet and adjust premiums to take that into consideration.

A remote survey of power-usage patterns could ascertain whether someone is at home, or is asleep. Unnoticed and quietly controlling the house, the smart technology could also report details of residents' activities to a third party whose purposes are hostile to their best interests.

Cautionary tales that transform the automated home into the haunted house exemplify broader worries about the lack of safeguards in current computing practice. Connecting appliances together in a standardized way and linking them to the Net offer innumerable opportunities for using information from a wide variety of sources. Without adequate safeguards, however, such connectivity risks random interference, counterproductive appliance behavior that is difficult to debug and control, and unwanted control and surveillance by computer systems representing other, potentially antagonistic interests.

These concerns reflect the Internet edge for networked home automation. They need to be addressed before home automation and connection to the Net can become widespread. How do we know we can trust our home automation systems? How do we know we can trust software and information that comes into a home system from the Net? How do we know who is using information gathered by sensors in a home automation system, or how it is being used? In this regard, the trusted system technologies discussed in chapter 3 could become as crucial for home automation as for digital money and digital publishing. For example, they could protect confidential information, reject unauthorized programming or control signals, and provide failsafe control policies.

Networked Design and Manufacturing

"You gotta be careful how you use a matter compiler, Nell."
Neal Stephenson, *The Diamond Age*

Over the past two hundred years, connection technologies have brought dramatic social and economic changes, shrinking distances and bringing together different communities and cultures. As we saw in the introduction, these technologies fall into two broad categories: transportation technologies like canals, railroads, automobiles, and interstate highways;

and communication technologies like the telegraph, telephone, radio, television, and the communication satellite. Transportation technologies move physical objects; communication technologies move information. Transportation carries mass; communication is massless. Transportation can move at the speed of the wind, but communications can move at the speed of light.

Distinctions between the two connection technologies fade in automated manufacturing: communication looks a lot like transportation if sending a message creates an object; and, in a limited way, fax machines and remote printers create a printed page. Stereolithography "prints" matter in three dimensions, causing physical objects to be constructed a layer at a time. A sender at one location transmits a set of manufacturing instructions telling a machine, a robot, where to put material, where to leave holes, and so on to create a physical object.

Digital Control of Manufacturing

Neal Stephenson's *The Diamond Age* describes an imagined future in which machines called *matter compilers* follow detailed instructions to assemble complicated objects according to a programmed design of atoms. The instructions could be sent over a network, so that, in effect, an object design created at one place can be transmitted to a distant matter compiler, which then assembles the object.

> The people here in Merkle Hall were all working on mass-market computer products . . . They worked in symbiosis with big software that handled repetitive aspects of the job. . . . An automated design system could always make something work by throwing more atoms at it. (Stephenson 1995)

Stereolithography is an analogous real technology that arranges plastic materials in arbitrary patterns to form an object. It produces three-dimensional parts from a computer description of a solid model without tooling, cutting, numerical control, or molding. It works with liquid plastics that harden when they are exposed to ultraviolet light. There are several variations on this technology. One approach lays down a plastic gel one layer at a time and then hardens it by exposing it to light from a laser. Another version uses a computer-controlled laser beam to draw cross sections of the object on the surface of a vat of liquid plastic. The plastic hardens only where it is touched by the laser, forming a thin cross-sectional layer of the part. The cross-section is then lowered by an elevator and the next layer is drawn by the laser, then adhered to the first section. This process is repeated until a complete solid plastic model is formed. Other related approaches to making solid models under computer control include selective laser sintering, fused deposition modeling, shell-production casting, topographic shell fabrication, and controlled milling. Laser sintering is a process of creating a solid mass, typically from a powdered material—plastics, ceramics, or metals—by selectively heating but not melting it.

The technologies for creating three-dimensional objects are evolving rapidly. Different manufacturing methods use materials varying in strength, flexibility, and transparency and work at different scales. Some methods can produce more accurate and predictable shapes than others, and some require later tooling to remove the scaffolding used at intermediate stages to provide temporary support.

Automated manufacturing technology goes back at least to the 1960s, when numerically controlled (NC) machines emerged as part of manufacturers' efforts to reduce costs in fabrication processes. As computers were very expensive at the time, the early machines typically received their instructions from computer-created control tapes. The machines had only a few sensors, no conditional control programs, and were relatively difficult to program. Each system had its own ad hoc control codes and programming conventions. Since then the industry has shifted to computer numerical control (CNC) and shop networks for interlinking machines. Computer-controlled manufacturing systems are now used worldwide by many producers, large and small. In the 1990s stereolithography for remote manufacturing attracted a lot of attention, but it is by no means the only technological option.

Edges for Automatic Manufacturing

From the perspective of magic and fairy tales, stereolithography and other kinds of numerically controlled manufacturing resemble genies. When we ask a genie to give us something, it uses magic to create the object. Other purveyors of magic in folk tales include tablecloths that instantly produce wonderful meals and empty bags that suddenly fill up with treasures. At the core of these tales is the dream of abundance for those who possess the magical appliance and say the magic words. Creating such capabilities through real advanced technologies has momentous social ramifications. Although there is little chance in the short term of their destabilizing economic systems or creating universal abundance, such technologies do raise issues about how we should organize the laws and economies related to the production and distribution of goods.

Patrick Salsbury imagines a time when people will exchange patterns or programs and pass gadgets around.

> Now think about being able to download the 3D data for just about anything, and having your computer "print" you a copy. New kitchen appliances, plates, cups, statues, candlesticks, lamp bases, bookends, whatever. . . . Right now, people are swapping software for free with each other. In a short time, people will be swapping gizmos with each other, saying "Look what I just made!" Shortly after that, everyone will be making things, and trading them around. (Salsbury 1997)

Besides the economic implausibility of creating home manufacturing systems in the near future, Salsbury's vision presents other, more complex questions. As it turns out, of course, most useful things are subtle and challenging to design; creating a workable design, even for such a mundane object as a bottle opener, is not like dashing off a quick note to a friend. Designers work hard on designs and protect that work by patents.

In a world of digitally generated designs, issues about protecting intellectual property arise, just as they do in book publishing. In our chapter 4 discussion of digital works at the Internet edge we considered several borderline cases, such as the protection of digital data compilations. In the case of computer-based designs for automatic manufacturing, there will probably also be borderline cases that raise questions about intellectual property. Is a given design novel enough or original enough to warrant legal protection? When building objects for automated manufacturing, should the designer just patent the designed object, or does its originality also lie in the program for building it? Can the program be protected by patent and copyright?

Our discussion of trusted systems for controlling the distribution and use of digital works also raises the question of whether programs for manufacturing digital objects can be considered digital works. In principle, agreements about their use could be written into digital contracts in the same way that a digital rights language sets forth the terms for using digital works. A consumer might, for example, purchase the right to assemble two copies of a wonderful bottle opener.

Finally, there are other potential edges to Salsbury's dream of a free-for-all in exchanged designs. With something like Stephenson's fictional Diamond Age nanotechnology, all sorts of objects could contain all sorts of hidden internal mechanisms. Like the stories of the Trojan horse or toy gremlins, seemingly harmless artifacts, downloaded free from the Internet, could wake up at midnight and perpetrate all kinds of mischief. Hackers now limited to making computer viruses could create real viruses or toys with bizarre or dangerous side effects. How would you know that a designed device was safe to manufacture?

Like so many other cases, Net connectedness moves us from a world in which we trust things we can see from people we know to one in which we use things we can't see from people we have never met. In principle, computers give us the opportunity to know more about many things than we do today and to document the conditions of trust. Perhaps a trusted system, analogous to the trusted systems for digital copyrights or the ones proposed for home automation will also repose at the heart of remote manufacturing devices. Such a trusted system could provide a formal means of assuring trust by checking the certificates and warrantees of items to be manufactured. As our social mindset moves from the local village to the global village, we will need to reinvent a whole new basis for trust.

Reflections

Western Machine technology . . . makes life seem easier, comfortable, cozy, but the price we pay includes the dehumanization of the self. To sleep in a cozy home, a good bed and eat great, chemically produced food you must rhyme your life with speed, rapid motion and time. The clock tells you everything and keeps you busy enough to forget that there could be another way of living your life.

Malidoma Somé, *Ritual: Power, Healing, and Community*

If sufficiently high technology is indistinguishable from magic—as suggested by Clarke, Newell, and many others—then the act of creating high technology puts magic into the world. Ironically, technology and machines are more often associated with the opposite point of view. Many writers argue that since the seventeenth century, technology and science have driven magic and superstition from the world. The increasing presence of advanced technology and the way new computer technologies are reshaping themselves in the image of magic offer us an opportunity to rethink that proposition.

Descartes and Huygens

In the absence of emotion and feeling, rationality breaks down.
Antonio Damasio, *Descartes' Error*

The idea that science and technology brought an end to the enchanted world is often linked with the work of René Descartes, the noted French philosopher and mathematician. Among others, Descartes argued that the sciences—by which the seventeenth century meant all knowledge—should be founded on certainty. There was considerable risk for this position. In 1633, Descartes suppressed publication of his work *The World* when he heard of Galileo's condemnation by the Inquisition for publishing similar ideas. Such misuse of power ultimately undermined the moral authority of the Church and contributed to the rise of scientific thought as an alternative, and less political, source of truth.

The rise of science also fostered a new attitude toward the universe, making it seem deterministic, a product of natural laws—essentially a machine. By framing the universe as fundamentally knowable, Descartes's science challenged its essential mystery and encouraged scientists to do experiments to discover its laws. Simultaneously, it disputed the privileged position of the clergy (or anyone else) in the search for knowledge.

Another seventeenth-century scientist, Christian Huygens, also contributed to the new view of the nature of the universe. Although he is most famous for grinding lenses, for his wave theory of light, and for his discovery of Saturn's rings, he was also the inventor of the pendulum clock. Clocks not only changed the way people regulated their days, they also became the machine people encountered most frequently during their lives. If the new worldview saw the universe as mechanized, clocks influenced this perception by providing a common example of the systematic and mindless nature of machines. As the universe was secularized, less room was left for God, mystery, or magic in everyday life.

In the late-twentieth century some scientists began to sense limits to how completely they could know and understand the universe—starting with the uncertainty principle and moving to other ideas about both the cosmos and the world of the infinitely small. Even if the universe is essentially a machine, it may have some counterintuitive properties and laws that place limits on what we can know about it. Some things may always contain mysteries.

Nonetheless, the scientific method continues to thrive as a way of learning about the universe.

In the Image of Magic

The ancient and modern magical technologies described in this chapter provide us with several clues to the underlying dreams and issues we face at the Internet edge. Wearable computers, like the magical clothing of folk tales, are intended to protect us from the harshness of the environment and to give us new powers, making us invulnerable to discomfort and inconvenience. Automated homes equipped with magical appliances are like the enchanted castle that—instead of equipping us to live in a harsh environment—changes the environment. Even automatic manufacturing systems, like Aladdin's genie, produce whatever we wish for.

Yet the darker side of magic revealed in cautionary tales is also instructive. Magical technologies—like Tolkien's rings or the red shoes—can have a will of their own. We could become too dependent on them and find to our horror that the enchanted castle has become the haunted house. The robots of automated manufacturing could get out of control, producing gremlins or running amok and burying us with things we don't need. We want the magic to control the environment; yet we fear that the magic will turn against us, bend us to its will, and diminish our selves and our world. Even our cars, the technological analogues of magic carpets, generate pollution as a by-product.

These images—both dreams and nightmares—are about power and control. All three technological versions of magic—the smart house, smart clothes, and smart manufacturing—emphasize our *separateness* from an environment perceived as inhospitable. Automated houses change the environment locally; because we can't change the environment everywhere, we wear smart clothes to give us portable protection; the natural environment does not provide us everything we think we need, so we use smart manufacturing to create it.

The danger is that the feelings of separateness from the environment induced by too much automation may foster a lack of care and respect for the earth. Our efforts to control the environment—to bend it to human will—can harm it and ultimately bring harm to ourselves.

We do not need to focus on wearable and networked computers to see the effects of mistreating the natural world. Since the early twentieth century the environmental movement has sought to bring attention to the results of clearing forests, overuse of pesticides, casual dumping of toxins, and overfishing our seas and rivers. A recurring effort of the movement has been raising the collective consciousness of the stakeholders. To naive consumers, food comes from grocery stores and restaurants. They pay money and get food; anything upstream of the point of purchase is out of sight and out of mind. From a food producer's point of view too, farming, cattle raising, and fishing practices are dictated solely by a need to be competitive and meet the consumer's desire for the least-expensive product. The problem with this perspective is that people are not just consumers and producers of food products. They are also breathers of air, drinkers of water, and beings who are themselves part of the environment. When the economic values of consumption and the marketplace dominate all other values untempered by conscious feedback about other values, the environment can be pushed seriously out of balance. The "unconscious machines" in this situation are not computers running amok but the aggregate effects of a society unconsciously acting contrary to its own best interests.

Thus, the stories of magic and the Internet edge technologies reveal a larger story about the exercise of power and unconscious action. An underlying conflict of values—between dominating and controlling the environment and caring for it—are at the center of most environmental and technological controversies.

A helpful insight from modern psychology—the concept of *projection*—can help explain the unconscious attribution of hidden personal qualities to other people and things. We can see such unconscious projections in our stories more clearly by mentally changing place with our machines and our environments. Are the stories about controlling nature or controlling ourselves? Do we need to tame the environment or tame our unconscious desire to control it? Is nature out of control and threatening to harm us, or are we out of control and threatening nature? Are the magical machines acting mindlessly, or are we? The point of these questions is not to assign credit or blame, or even to find a single answer. Rather, it is to make us conscious of projections we may be unaware of.

Flight from technology and the return to nature is a classic utopian theme. In various ways, the utopian literature—and many of us personally—is asking "What is missing and what are we trying to achieve?" Perhaps we need to find a way to reframe our understanding of what is missing. We may discover that "it" is not "out there" in the world but, rather, is something inside our selves and in the world we create. As the Internet edge about magic and machines reflects a deeper issue of power and control, it gives us an opportunity to consider carefully what we are creating. In shaping the Net, our computers, and our environment, we are also shaping ourselves.

There are always tensions and confusions at an edge. Moving through an edge requires us to synthesize a course of action from the opposites in tension. We can consider not only what opportunities the Net offers for taking care of us but what it offers for taking care of the earth. What are the opportunities for bringing greater thoughtfulness and mindfulness to our machines as well as to ourselves?

Technology and Mindfulness

One of the elders asked, "Where do these white people run to every morning?"
"To their workplaces, of course."
"Why do they have to run to something that is not running away from them?"
"They do not have time." I had to say this word in French because there is no equivalent in the local language. The conversation came to a halt when the elder had to ask what this "time" was.

Malidoma Somé, *Ritual: Power, Healing, and Community*

When the elder asked where the people were running to, he was indirectly pointing out that moving quickly can also serve as a distraction. When things move too fast, we have difficulty paying attention to them. When we slow down, we discover things we would otherwise not see. Slowing down gives us the time to focus, withdrawing our minds from some things in order to pay attention to others.

David Levy, a computer scientist and teacher of mindfulness, believes that American middle-class culture (and perhaps other cultures as well) is characterized by increasing levels of fragmentation: "Being too busy is one of the most common complaints of our times. . . . There is a sense of being fractured, incomplete, less than whole" (Levy 1997). Levy credits James Beniger's book *The Control Revolution* (1989) with providing insights into the speedup he sees as the cause of fragmentation. Beniger's historical account begins with the invention of steam power in the late-eighteenth century. In the nineteenth century, steam powered railroads, ocean-going vessels, and many manufacturing plants, increasing the rate at which raw materials could be gathered and consumed. As organizations became bigger, faster, and more widely distributed, systematic management methods were introduced to extend centralized control. Larger organizations required new information technologies to manage the increasingly complex processes of production and distribution. According to Beniger, the new economic engine could produce goods faster than people could consume them; so modern advertising methods were invented to create new needs and promote higher consumption. In this way, steam power, information management, and advertising joined together to accelerate the pace of modern Western life—perhaps leading Somé's elder to wonder where everyone is running.

Although Beniger's account somewhat oversimplifies the history of the modern era, it raises pertinent questions about the effect of technologies. The safeguards built into the SkyTel system failed to cope well enough to prevent beepermania and the risks associated with fan-out. This episode and, more generally, the evolution of beepers and cellular telephones during the late 1990s, illuminates a fundamental tension in modern living: balancing our need for accessibility with the desire for uninterrupted personal time.

Connection technologies are not static, however; beepers and cell phones are continuing to evolve to fit into our lives better. The second generation of beepers features quiet vibrational signals, so that annoying beeps do not disturb everyone near users. Phone controllers can be programmed with caller-identification signals and passwords to screen calls. Even without such technological innovations, we can mediate the relative priorities of being accessible and being alone by simply turning off the cell phone and letting our calls be transferred to voice mail.

The edge work related to wearable computers is just beginning to unfold. In principle, wearable computers can put us in touch with information and regions that are far from our visible environment. At the same time, they can isolate us and distract our attention from crucial parts of our immediate environment. Which is the greater benefit? Which the greater risk? The isolation that leads to a sense of personal separateness or the telepresence that can foster a sense of social connection? The distraction that leads to fragmentation of our consciousness or timely information that brings us information in context when we need it? The answers may lie in the skill we and others bring to designing such technologies and the wisdom we apply to deciding when to use them.

From a perspective of environmental consciousness, the Net may even provide an antidote to unconscious action at a distance. When a baby moves its hand and sees its hand moving, it develops a sense of self that identifies the hand as a part of "self" and identifies a toy, for example, as a part of "not-self." By analogy, if we could more readily see or feel the consequences of our actions on the environment, we might feel less isolated and separate from it. If the Net could show us readily and immediately the effects of our choices, even at a distance, it might counteract the illusion that what we do is not important in the long run.

The senses and nervous systems of our bodies tell us when we touch something hot, warning us before we hurt ourselves. At the end of the nineteenth century, novelist Nathaniel Hawthorne likened the telegraph to the new "electronic nervous system" of the planet. The Internet is potentially a much more powerful instrument than the telegraph for providing sensory information from a distance and effecting remote feedback and action. This is not to say that simply having more communications technology will, by itself, lead to solutions for environmental problems. After all, we already have telephones, magazines, newspapers, televisions, fax machines, and the post office. Oil spills not withstanding, major environmental disasters seldom develop overnight. The challenge we face is to find ways to use our communication and information capabilities to take more direct responsibility for noticing and regulating the effects of our actions on the world and for recognizing that actions at one place have increasingly noticeable effects far away.

In his famous silent movie *Modern Times,* Charlie Chaplin struggles against the machinery of a manufacturing plant. The machinery and the society around it are racing forward mindlessly, perhaps, eventually, to spin out of control. The opportunity for us, in creating the Net, is to build a technology that gives us the space not only to move quickly but also to notice where we are, to harness natural energies but also to do so with care and deliberation. Ultimately, the choices are ours. What will survive into the next generation depends on what we value. Rather than letting technology lead to the fragmentation of our lives and consciousness, we have to rise to the challenge of wholeness to make choices that enhance our whole human development.

Epilogue: The Next Edge and Discovering Ourselves

In bringing closure to writing this book, two main questions arise for me: "What comes next?" and "What does it all mean?"

More Waves from the Net

What is the next edge? In collective edges like the Internet edge there is no single answer. Nonetheless, it seems appropriate to speculate about edges we may come to in the future. In that spirit, I return to the trends we looked at in the last chapter.

The trend toward networking devices begun in digital manufacturing is being applied in more and more areas of production. This is true not just of the relatively exotic technology of stereolithography, but also of much more ordinary manufacturing processes—drilling of materials; folding, cutting, or milling metals; printing on various materials; and sewing and weaving. Networking makes it possible for people to sell time on their manufacturing machines through the Net and to receive digital instructions for making objects.

This is not to say that there will be such a renaissance of interest in making things that everyone will be doing it. It is more likely that specialists in designing and programming for automated manufacturing will advertise their services on the Net; others who want to have something custom made might also advertise their needs. Web-based auctions, already used in a small way, could connect people having unusual requests with experts whose special skills and excess manufacturing capacity could fulfill them. The "Net effect" of all this may be waves of unique garments and extraordinary gadgets designed for special interest groups. Because much of the early work and design would be done invisibly by people at scattered locations on the Net, these waves of items would appear to spring up everywhere at the same time.

Imagine teenagers marking their "clans" with distinctive kinds of clothing or jewelry not available through the usual retail channels. Imagine sports enthusiasts—say bicyclists or surfboarders—suddenly appearing with unusual new kinds of gear. Imagine art objects or car accessories s that appear almost overnight, seemingly out of nowhere. The Net could trigger waves of cultural experimentation in people and groups who are invisible until they show up with their unique and possibly weird artifacts. The Net could trigger further reorganizations of business and manufacturing, with outsourcing of design, manufacturing, and distribution. This is expected to unfold over the next twenty years.

Discovering Who We Are

Q. What is necessary?
M. To grow is necessary. To outgrow is necessary. To leave behind the good for the sake of the better is necessary.
Sri Nisagadatta Maharaj, *I Am That*

Having now at least delivered one answer to the "What's next?" question—in terms of the growth of radically-distributed manufacturing—we turn to the other question, "What does it all mean?"

An interesting property about edges is that new ones keep coming. When we are working on one edge—starting, peaking, and trailing—the next one is beginning to take shape. Edges are like waves on the ocean, but more varied. Each one prepares the way for the next.

If we imagine ourselves to be swimmers amid all these waves of change, what is enduring in all this? It is in this spirit that I approach the question about meaning.

Bhagavan Ramana Maharsi, the great Indian sage, was known for his simple teachings. When he taught he generally asked his visitors to ask themselves just one question: "Who am I?" His question, like the Western admonition to "know thyself," is asked in some form in most spiritual traditions. Sri Nisargadatta Maharaj observes that when one meditates on this question, looking very carefully at the answers that come to mind, the self becomes hard to pin down. We are more than our positions in our families, more than our occupations, more than our age group, more than our particular interests, more than our nationalities, more than our races. Even our thoughts come and go. So what is the enduring part of who we are? The ego fades away as the boundary between the self and the other fades away.

It seems to me that the Internet edge raises a collective analogue to this question. *Who are we?* By connecting our world cultures more closely together, the Net brings us face to face with our differences and similarities. It challenges diversity. It triggers waves of cultural mixing as the Net crosses boundaries. Like the next edge, it may trigger new waves of cultural exploration, unleashing forces that simultaneously diminish cultural diversity and create new cultural diversity. Either way, it challenges the status quo of who we are.

Reflections on the Mountain

The birds have vanished into the sky,
and now the last cloud drains away.
We sit together, the mountain and me,
until only the mountain remains.

Li Po

And then the mountain disappears
without a trace
And all it took
Was a sudden Leap of Faith.

Kenny Loggins and Guy Thomas, from the *Leap of Faith* album, Columbia Records

Mountains are special in many spiritual practices. In the winter, snowstorms may swirl around the mountain. But when the storm clears, the mountain is still there. In the summer, people may swarm over the mountain making picnics and hiking to the summit. And in the evening, the silence of the mountain is still there. The mountain symbolizes an eternal core and presence, impervious to commotion and change.

In Tibet, where I trekked during the last phase of writing this book, there is a holy mountain known as Mount Kailash. With us on our trek were three Nepalese shamans, who are expected to come to the mountain at least three times during their lives. They come once when they receive their calling, once in mid-life to renew their power and purpose, and once at the end of life, to release their spirit. When shamans die before making this last journey, their villages have an obligation to perform a ritual for returning the spirit to the mountain.

For the Buddhists, the mountain is also holy and is said to be the center of the world. A great festival is held there in June to celebrate the birthday of Buddha's enlightenment. Buddhists are expected to walk around the mountain in the clockwise direction. A more ancient, indigenous Tibetan religion Bön, practiced by the Bön Po, also regards the mountain as sacred. Its religious symbols are often similar to those of Buddhism but reversed in some way; they walk around Kailash in the counterclockwise direction. On our trek we encountered and met with people of all of these spiritual orders, each of them honoring the mountain in their own way.

For Hindus as well, Mount Kailash is a holy mountain, the home—and also the embodiment—of Shiva. Within all the variations and manifestations of Shiva are the essence of three faces—creating, sustaining, and destroying. As we saw in earlier chapters (especially chapter 7), these faces are also characteristic of the stages of an edge. Edges mark a change. A new order is created, sustained, and finally passes. The faces are metaphors for life's transformations. Life is full of change.

It is paradoxical that Mount Kailash should be said to embody Shiva, for the mountain that represents the god of transformation also embodies steadiness and changelessness. The mystery within that contradiction brought me to select the two short poems. The first, by the Chinese philosopher Li Po, speaks of the transitory nature of life, as compared with the timeless nature of the mountain. The second, from a contemporary popular song, speaks of faith. In this excerpt, it is the mountain that disappears. What is timeless is not physical but is part of what some would call our true nature.

When we see the many different faces of each other, may we recognize them as the faces of our selves. In this way the Net as a change amplifier and the Net as a portal into cyberspace and the Net as a knowledge medium can also be an agent that moves us further into the mystery of life and deeper into an understanding of who we are.

About the Author

Mark Stefik is a principal scientist at the Xerox Palo Alto Research Center. During the writing of this book he was manager of the Secure Document Systems area in the Computer Science Laboratory, and manager of the Human-Documents Interaction area in the Information Sciences and Technologies Laboratory. He became interested in how things work and understanding what matters during his secondary education at Benson Polytechnic in Portland, Oregon. At Stanford University, he earned a bachelor's degree in mathematics and a doctorate in computer science.

Stefik's early research was in hierarchical planning and design in the areas of molecular genetics and VLSI design. He has carried out research in object-oriented programming languages, computer support for collaborative work, and trusted systems for digital property rights. A recurring theme of Stefik's research and writing is using technology to enhance creativity, collaboration, and self-expression. He is the author of *Introduction to Knowledge Systems* (Morgan Kaufmann 1995) and *Internet Dreams: Archetypes, Myths, and Metaphors* (MIT Press 1996). He was elected a fellow of the American Association for Artificial Intelligence in 1990 and of the American Association for the Advancement of Science in 1997. His electronic-mail address is stefik@parc.xerox.com.

References

Agre, Philip E., and Rotenberg, Marc, eds. 1997. *Technology and Privacy: The New Landscape*. Cambridge: MIT Press.

Anderson, Ross, and Kuhn, Markus. 1996. Tamper resistance—A cautionary note. *Proceedings of the Second Workshop on Electronic Commerce*, Oakland, Calif., November 18–20. Also available online at http://www.cl.cam.ac.uk/users/rja14/tamper.html.

Apple Computer, Inc. v. *Microsoft Corp.*, 35 F.3d 1435 (9th Cir. 1994).

Atari GamesCorp. v. *Nintendo of America, Inc.* 975 F.2d 832 (Fed Cir. 1992).

Baker, Kim. 1998. Quoted in Betsy Schiffman, "On Her Own." *Palo Alto Weekly*, February 13.

Barlow, John Perry. 1994. "The Economy of Ideas." *Wired (*March): 85.

Bellotti, Victoria. 1997. Design for privacy in multimedia computing and communications environments. In Agre and Rotenberg 1997.

Beniger, James. 1989. *The Control Revolution. Technological and Economic Origins of the Information Society*. Cambridge: Harvard University Press.

Bennett, Colin J. 1997. Convergence revisited: Toward a global policy for the protection of personal data. In Agre and Rotenberg 1997.

Berle, Adolf A. 1969. *Power*. New York: Harcourt Brace & World.

Bettelheim, Bruno. 1977. *The Uses of Enchantment: The Meaning and Importance of Fairy Tales*. New York: Alfred A. Knopf.

Bush, Vannevar. 1945."As We May Think," *Atlantic Monthly* (July). (An excerpt reprinted in Stefik 1996.) Also available on-line at www.isg.sfu.ca/~duchier/misc/vbush/.

Buxton, Bill. 1997. At Seminar on People, Computers, and Design, November 21. Available on-line at http://pcd.stanford.educ/old/html/971121-buxton.html.

Campbell, Joseph, with Moyers, Bill. 1988. *The Power of Myth*. New York: Doubleday.

Card, Stuart. 1997. Personal communication, June.

Caruso, Denise. 1997. "Digital Commerce: The Interactive Media Industry Begins to Deconstruct Its Self-made Myths." *New York Times*, April 7, p. D7. Available on-line at http://www2.nando.net/newsroom/ntn/info/040797/info1_4958.html.

Clarke, Arthur C. 1968. *2001: A Space Odyssey*. London: Hutchinson/Star. (A novel based on the original screenplay by Stanley Kubrick and Arthur C. Clarke)

Clinton, William J., and Gore. Albert. 1993. *A Framework for Global Electronic Commerce*. Available on-line at http://www.iitf.nist.gov/eleccomm/ecomm.html.

Computer Associates International, Inc. v. Altai, Inc., 982 F.2d 693 (2nd Cir. 1992).

Copyright Law of 1976. *United States Code*, title 17, sections 108–120.

Crichton, Michael. 1995 *The Lost World: A Novel*. New York: Knopf.

Crigler, Jeff. 1997. Oral interview, February 19.

Damasio, Antonio R. 1995. *Descartes' Error: Emotion, Reason, and the Human Brain*. New York: Avon Books.

Dawkins, Richard. 1990.*The Selfish Gene,* 2nd ed. New York: Oxford University Press.

Department of Defense 1985. *Department of Defense Standard: Trusted Computer System Evaluation Criteria*. DOD 5200.28-STD (Orange Book), December 26. Also available at http://www.radium.ncsc.mil/tpep/library/rainbow/5200.28-STD.html/.

Estés, Clarissa Pinkola. 1992.*Women Who Run With the Wolves: Myths and Stories of the Wild Woman Archetype*. New York: Ballantine Books.

European Parliament and Council of Europe. Directive on the Protection of Personal Data (95/46/EC). 1995. It is available on-line on-line at http://jilt.law.strath.ac.uk/jilt/dp/material/directiv.htm.

Feist Publications v. *Rural Telephone Service*, 499 U.S. 340 (1991).

Gold, Rich. 1997. No information without representation. E-mail message, November 27.

Goldstein, Ira, and Papert, Seymour 1977. Artificial intelligence, language, and the study of knowledge. *Cognitive Science* 1, no. 1: 84–123.

Grove, Andrew S. 1996. *Only the Paranoid Survive: How to Exploit the Crisis Points that Challenge Every Company and Career.* New York: Currency/Doubleday.

Huberman, Bernardo A., and Hogg, Tadd. 1987. Phase transitions in artificial intelligence systems. *Artificial Intelligence* 33:155–71.

Kabat-Zinn, Jon. 1994. *Wherever You Go, There You Are: Mindfulness Meditation in Everyday Life*. New York: Hyperion, 1994.

Kahin, Brian, and Nesson, Charles, eds. 1997. *Borders in Cyberspace: Information Policy and the Global Information Infrastructure*. Cambridge: MIT Press.

Kaplan, Jerry. 1995. *Startup: A Silicon Valley Adventure Story*. Boston: Houghton Mifflin Co.

Kedzie, Christopher R. 1997. The third waves. In Kahin and Nesson 1997: 106–28.

Kelly, Kevin T. 1995. *Out of Control: The New Biology of Machines, Social Systems and the Economic World.* Reading, Mass.: Addison-Wesley.

Kinoshita, Hiroo.1985. Towards an advanced information society. Address to Second Institute for New Generation Computer Technology. *ICOT Journal*, no. 6 (February).

Kirkpatrick, John, and Sandberg, Jared. 1997. *Wall Street Journal.* January 10, p. 1.

Lane, Carole. 1997. *Naked in Cyberspace: How to Find Personal Information Online.* Wilton, Conn.: Pemberton Press.

Lavendel, Giuliana. 1997. Oral interview, February 19.

Lederberg, Joshua 1997. Oral interview, June 25.

Lenat, Douglas B. 1995. CYC: A large-scale investment in knowledge infrastructure. *Communications of the ACM* 38, no.11 (November): 33–38.

Levy, David 1997. Personal communication.

Licklider, J.C.R. 1965. *Libraries of the Future.* Cambridge: MIT Press. (Excerpt reprinted in Stefik 1996.)

Li Po. Quoted in Kabat-Zinn 1994.

Loggins, Kenny, and Thomas, Guy. 1991. "Leap of Faith." In *Leap of Faith* album, Columbia Records. Used by permission [?] of Gnossos Music. ASCAP/Windham Hill Music (BMI).

Maharaj, Sri Nisagadatta. 1973. *I Am That: Talks with Sri Nisagadatta Maharaj,* Maurice Frydman, transl., and Sudhakar S Dikshit, ed. Durham, N.C.: Acorn Press.

Mann, Steve. 1996. 'Smart clothing': Wearable multimedia computing and 'personal imaging' to restore the technological balance between people and their environments. *Proceedings of ACM Multimedia* '96, pp. 163–74.

Mayer-Schönberger, Viktor. 1997. Generational development of data protection in Europe. In Agre and Rotenberg 1997.

Michie, D. 1984. A prototype knowledge refinery. *In Intelligent Systems: The Unprecedented Opportunity,* J. E. Hayes and D. Michie, eds. Chichester: Ellis Horwood.

Miller, Matthew. 1997. Personal communication.

Mills, Dick. 1998. Personal communication.

Mindell, Arnold, and Mindell, Amy. 1991. *Riding the Horse Backwards: Process Work in Theory and Practice.* London: Penguin Books.

Moore, Geoffrey A. 1991. *Crossing the Chasm: Marketing and Selling Technology Products to Mainstream Customers.* New York: Harper Business.

Moore, Geoffrey A. 1995. *Inside the Tornado. Marketing Strategies from Silicon Valley's Cutting Edge.* New York: Harper Business.

Moore, Robert L., and Gillette, Douglas. 1990. *King, Warrior, Magician, Lover: Rediscovering the Archetypes of the Mature Masculine.* San Francisco: HarperCollins.

Moore, Thomas. 1992. *Care of the Soul: A Guide for Cultivating Depth and Sacredness in Everyday Life*. New York: HarperCollins.

National Information Infrastructure Task Force. 1997. *Options for Promoting Privacy on the National Information Infrastructure*. It is available on-line at http://www.iitf.nist.gov/ipc/privacy.html.

Network Business Services web page: http://www.nim.com.au/inet_pub/in04001.html

Newell, Allen. 1992. Fairy tales. *AI Magazine* 13, no. 4 (Winter):46–48.

Newman, Stagg. 1997. Address on design of the National Information Infrastructure (NIII). In Hundt, Reed. ed. 1997. *Federal Communications Commission Bandwidth Forum*, January 23. Available on-line at http://www.fcc.gov/Reports/970123.txt.

Nunberg, Geoff. 1997. Interview on National Public Radio. *Fresh Air*, November 4.

Pappas, Charles. 1997. To surf and protect. *ZD Net* (December). Available on-line at http://www.zdnet.com/yil/content/mag/9712/pappas9712.html.

Pitroda, Sam. 1993. Development, democracy, and the village telephone. *Harvard Business Review* (November-December).

Pool, Ithiel De Sola. 1983. *Technologies of Freedom*. Cambridge: Belknap Press of Harvard University Press.

Price, Derek de Solla. 1983. "Sealing Wax and String: A Philosophy of the Experimenter's Craft and Its Role in the Genesis of High Technology." Address to the AAAS Convention, Detroit. Washington, D.C.: American Association for the Advancement of Science.

ProCD, Inc. v. *Zeidenberg*, 86 F.3d 1447 (7th Cir.).

Risher, Carol. 1980–1985. "Technology Watch." In *AAP Newsletter* (Association of American Publishers). February 1980–October/November 1985.

Salsbury, Patrick. 1997. Web page on stereolithography. http://reality.sculptors.com/stereolithography.html.

Samuelson, Pamela. 1996a. Regulation of technologies to protect copyrighted works. *Communications of the ACM* (Association for Computing Machinery) 39, no. 7 (July): 17–22.

Samuelson, Pamela. 1996b. On author's rights in cyberspace: Are new international rules needed? *First Monday*, October. Available on-line at http://www.firstmonday.dk/issues/issue4/samuelson.

Sarpangal, Mali. 1997. Oral interview.

Schilit, William, Golovchinsky, G., and Price, Morgan N. 1998. Beyond paper: Supporting active reading with free-form digital ink annotations. In *Proceedings of CHI 98 Conference* (Computers and Human Interaction), pp. 249–56. New York: ACM Press.

Schrage, Michael. 1993. "Innovation: The Day You Discover That Your House Is Smarter Than You Are." *Los Angeles Times*, November 25.

Sega Enterprises, Ltd. v. *Accolade, Inc.*, 977 F.2d 1510 (9th Cir. 1993).

Shaw, Ronald E. 1966, 1990. *Erie Water West: A History of the Erie Canal, 1792–1854*. Lexington: University Press of Kentucky.

Simon, Herbert A. 1997. The future of information systems. *Annals of Operations Research*. 71:3–14.

Somé, Malidoma Patrice. 1993. *Ritual: Power, Healing, and Community*. Portland, Ore.: Swan/Raven.

Sony Corp. of America v. *Universal City Studios, Inc.*, 464 U.S. 417 (1984).

The Source. 1980, 1981. *Users' Manual & Master Index to the Source*.

Spencer, Paula Underwood. 1990. "A Native American Worldview." *Noetic Sciences* (Summer 1990): 102–105.

Stefik, Mark. 1986. The next knowledge medium. *AI Magazine* 7, no. 1 (Spring).

Stefik, Mark. 1995. *Introduction to Knowledge Systems*. San Francisco: Morgan Kaufmann.

Stefik, Mark, ed. 1996. *Internet Dreams: Archetypes, Myths, and Metaphors*. Cambridge: MIT Press.

Stefik, Morgan. 1997. Personal communication via e-mail.

Steiger, Bettie. 1998. Oral interview, January.

Stephenson, Neal. 1995. *The Diamond Age, or a Young Lady's Illustrated Primer*. New York: Bantam Books.

Toffler, Alvin. *Future Shock*. 1970. New York: Random House.

Tolkien, J.R.R. 1977. *The Silmarillion,* Christopher Tolkien, ed. Boston: Houghton Mifflin.

Ueland, Brenda. 1938. *If You Want to Write: A Book about Art, Independence and Spirit*. New York: G.P. Putnam's Sons. Paperback reprint, St. Paul, Minn.: Graywolf Press, 1997.

Varian, Hal. 1995. *Scientific American* (September): 200.

Volkmer, Ingrid. 1997. Universalism and particularism: The problem of cultural sovereignty and global information flow. In Kahin and Nesson. 1997.

von Franz, Marie-Louise. 1972. *Problems of the Feminine in Fairytales*. Irving, Texas: Spring Publications.

Walker, Barbara K., and Walker, Warren S., eds. 1963. *The Erie Canal: Gateway to Empire*. Boston: D.C. Heath.

Weber, Eugen J. 1976. *Peasants into Frenchmen: The Modernization of Rural France, 1879–1914*. Stanford, Calif.: Stanford University Press.

Weiser, Mark. The computer for the twenty-first century. 1991. *Scientific American* (September): 94–104.

Wilber, Ken. 1979. *No Boundary: Eastern and Western Approaches to Personal Growth*. Los Angeles: Center Publications.

Woodman, Marion, and Dickson, Elinor. 1996. *Dancing in the Flames: The Dark Goddess in the Transformation of Consciousness*. Boston: Shambhala.

Suggestions for Further Reading

Here are some suggested readings for readers interested in pursuing topics in greater depth. The readings range from technical to general sources and include selected primary documents as well as technical journals, books, and relatively ephemeral web sources. The goal of this listing is to provide diverse and accessible starting places and a few gems rather than a comprehensive bibliography. The readings are arranged topically by chapter and do not duplicate items cited in the Reference list.

Chapter 1: The Internet Edge

Friedlander, Amy. 1995. *Natural Monopoly and Universal Service: Telephones and Telegraphs in the U.S. Communications Infrastructure*, 1837–1940. Reston, Va.: Corporation for National Research Initiatives.

Chapter 2: The Portable Network

Advanced Television Systems Committee. Home Page: http://www.atsc.org/index.html. Information on the DTV bandwidth allocations for digital television and other services are described in a final report available on-line at http://www.atsc.org/acats.html.

AER Energy Resources Home Page. It includes several on-line descriptions of battery technology, especially zinc-air batteries. See http://www.aern.com/zincair.html.

Alles, Anthony. ATM Internetworking. Available on-line at http://cell-relay.indiana.edu/cell-relay/docs/ftp.cisco.com/ATM-Internetworking.pdf.

Cisco Systems. *Internetworking Terms and Acronyms.* Available on-line as http://www.pluscom.ru/CD–20/data/doc/cintrnet/ita.htm.

Current On-line Briefing—Public TV and the transition to digital broadcasting: http//www.current.org/atv1.html#Links

Das, Sajal K., Sanjoy, K. Sen, and Jayaram, Rajeev. 1997. A dynamic load balancing strategy for channel assignment using selective borrowing in cellular mobile environment. *Wireless Networks* 3, no. 5 (October): 333–47. (This professional journal covers topics in mobile communication, computation, and information.)

dpiX Home Page. Information about the company's dpiX displays available on-line at http://www.dpix.com.

Federal Communications Commission site: http://www.fcc.gov. For a copy of the rules for Local Multipoint Distribution Service, see http://www.fcc.gov/Bureaus/Wireless/Orders/1997/fcc97082.txt.

Fujitsu Corporation. For information on their pen computers, see http://www.fpsi.fujitsu.com/.

Horneffer, M., and Plassman, D. Directed antennas in the Mobile Broadband System. Available on-line as http://www.comnets.rwth-aachen.de/~maho/paper/paper.html.

Jenkins, Reese V., Rosenberg, Robert A., et al. eds. 1989–1995. *The Papers of Thomas A. Edison*, 3 vols. Baltimore: Johns Hopkins University Press.

Kanellos, Michael. 1997. Moore says Moore's Law to hit wall. c|net News.com. September 30, Available on-line at http://www.news.com/News/Item/0%2C4%2C14751%2C00.html?nd.

Lorch, Jacob R., and Smith, Alan Jay. 1997. Scheduling techniques for reducing processor energy use in MacOS. *Wireless Networks* 3, no. 5 (October): 311–24.

McAuliffe, Kathleen. 1995. The undiscovered world of Thomas Edison. *Atlantic Monthly* (December): 80–93.

National Information Infrastructure 2000 Steering Committee. 1996. *The Unpredictable Certainty: Information Infrastructure through 2000*. Computer Science and Telecommunications Board, National Research Council. Washington, D.C.: National Academy Press.

NEC Corporation Home Page contains information about its NEC displays. http://www.nec.com

Nikeii. 1997. The uphill battle of 40-inch plasma displays. *Nikkei Electronics Asia* 6, no 7 (July). Available on-line at http://www.nikkeibp.com/nea/july97/specrep/index.html.

PC Systems Home Page includes information on their handheld computers for data entry applications. PC is a reseller of hardware by Symbol, Epson, and Fujitsu. http://www.pcsystems-usa.com/index.html

Richter, W. D. (director), and Rauch, Earl Mac (scriptwriter). 1984. *The Adventures of Buckaroo Banzai Across the 8th Dimension*. The film script is available on-line at http://jerseyguy.com/banzai.html and at http://kumo.swcp.com/synth/text/buckaroo_banzai_script.

Ricochet Corporate Home Page. http://www.ricochet.net

Rout, L. 1997. Telecommunications: The spoils of war. *Wall Street Journal* (special report), September 11.

Schilit, B., Douglas, F., Kristol, David M., Krzyzanowski, P., Sienicki, J., and Trotter, J. 1996. TeleWeb: Loosely connected access to the World Wide Web.

Computer Networks and ISDN Systems 28: 1431–44. Available on-line at http://www5conf.inria.fr/fich_html/papers/P47/Overview.html.

Sematech. 1997. *The National Technology Roadmap for Semiconductors.* Available on-line at http://www.sematech.org.

Service, Robert E. 1996. Can chip devices keep shrinking? *Science* 274 (December): 1834–36.

Time Domain. Technical papers on impulse radio. http://www.time-domain.com/technology.html

Universal ADSL Working Group Home Page. http://www.uawg.org/index.html

URLs for the History of Inventions

There are many on-line sources of information about inventors and the early history of the sound recording industry and radio.

Arthur C. Clarke Foundation. An on-line history of communication. http://www2.acclarke.co.uk/acclarke

Edison National Historic Site. Historical materials about Thomas Edison and the National Phonograph Company. http://www.nps.gov/edis/ed724700.htm

MIT web site. General catalog of inventors. http://web.mit.edu/invent/

National Inventors Hall of Fame. General information about American inventors. http://www.invent.org/book/index.html

National Library of Canada. A page about Emile Berliner and the museum dedicated to his work. http://www.nlc-bnc.ca/services/eberliner.htm

Rutgers University. A catalog of on-line papers about the life, inventions, and business of Thomas Alva Edison. http://edison.rutgers.edu/taep.htm

In addition to these general information sites, there are a number of on-line information pages about the corporate history of companies presently in the music recording and office dictation businesses. These include:

Sony. History of Columbia Records. http://www.music.sony.com/Music/PreswsInfohist.html.

RCA Victor. History of RCA. http://www.rca-electronics.com/story/later.

Dictaphone Corporation. History of Columbia Graphophone. http://ourworld.compuserve.com/homepages/dictaphone/history.htm.

A European perspective is available from EMI at http://www.emiclassics.com/centenary/history.html.

Chapter 3: The Digital Wallet and the Copyright Box

Aucsmith, David. 1996. Tamper-resistant software: An implementation. In Ross Anderson, ed., *Information Hiding: Proceedings of First International Workshop, May 1996*. Berlin: Springer.

Clark, Don. 1997. "Site Unseen: Facing Early Losses, Some Web Publishers Begin to Pull the Plug." *Wall Street Journal*, January 14, p A20.

Fukuyama, Francis. 1995. *Trust: The Social Virtues and the Creation of Prosperity*. New York: Free Press.

Kahin, Brian, and Arms, Kate, eds. 1996. Forum on technology-based intellectual property management. *Electronic Commerce for Content: A Journal of the Interactive Multimedia Association* 2 (August).

Kent, S. T. 1980. Protecting externally supplied software in small computers. Ph.D. diss., MIT (MIT/LCS/TR–255).

Pfleeger, Charles P. 1989. *Security in Computing*. Prentice Hall, Englewood Cliffs, N.J.

Summers, Rita C. 1997. *Secure Computing: Threats and Safeguards*. New York: McGraw-Hill.

Tygar, J. D., and Yee, B. 1994. Dyad: A system for using physically secure coprocessors. *Technological Strategies for Protecting Intellectual Property in the Networked Multimedia Environment* 1, no 1 (January):121–52. Also available from Interactive Multimedia Association at http://www.cni.org/docs/ima.ip-workshop/www/Tygar.Yee.html.

White, Steve R., and Comerford, Liam. 1990. ABYSS: An architecture for software protection. *IEEE Trans. on Software Engineering* 16, no. 6 (June).

Chapter 4: The Bit and the Pendulum

American Law Institute and National Conference of Commissioners on Uniform State Laws. 1997. Uniform Commercial Code [UCC] Article 2B - Licenses. Draft of May 5 Available at <http://www.law.uh.edu/ucc2b/050597/0505_2b.html>. Draft of 1998 proposed revision is available from the UCC Article 2B Revision Home Page at <http://www.law.uh.edu/ucc2b/>, or from the National Conference of Commissioners on Uniform State Laws, Drafts of Uniform and Model Acts, official site at <http://www.law.upenn.edu/library/ulc/ulc.htm>.

Cohen, Julie E. 1996. A right to read anonymously: A closer look at copyright management in cyberspace. *Connecticut Law Review* 28, no. 4 (Summer): 981–1039.

Elkin-Koren, Niva. 1997. Copyright policy and the limits of freedom of contract. *Berkeley Technology Law Journal* 12, no. 1: 93–113.

Lemley, Mark A. 1995. Intellectual property and shrinkwrap licenses. *Southern California Law Review* 68: 1239.

Merges, Robert P. 1997. The end of friction? Property rights and contract in the "Newtonian" world of on-line commerce. *Berkeley Technology Law Journal* 12, no.1: 115–36.

Nelson, Ted. 1981. *Literary Machines*. (Copies available from Ted Nelson, Box 128, Swarthmore, Pa. 19081.)

O'Rourke, Maureen A. 1997. Copyright preemption after the ProCD case: A market-based approach. *Berkeley Technology Law Journal* 12, no.1: 53–91.

Samuelson, Pamela. 1999. Good news and bad news on the intellectual property front. *Communications of the ACM* 2, no. 3 (March): 19–24.

Schlachter, Eric. 1997. The intellectual property renaissance in cyberspace: Why copyright law could be unimportant on the Internet. *Berkeley Technology Law Journal* 12, no.1: 15–51.

Stanford University Libraries. Copyright and fair use page. <http://fairuse.stanford.edu/>

Stefik, Mark. 1996. Letting loose the light: Igniting commerce in electronic publication. In Stefik 1996.

Stefik, Mark. 1997. Trusted systems. *Scientific American* (March): 78–81, Also available on-line at http://www.sciam.com/0397issue/0397stefik.html.

Stefik, Mark. 1997. Shifting the possible: How trusted systems and digital property rights challenge us to rethink digital publishing. *Berkeley Technology Law Journal* 12, no.1: 137–59.

U.S. Department of Commerce, Information Infrastructure Task Force. 1995. *Intellectual Property and the National Information Infrastructure: The Report of the Working Group on Intellectual Property.* Available at <http://www.uspto.gov/web/offices/com/doc/ipnii>.

Chapter 5: Focusing the Light

Baldonado, Michelle, Wang, Q, and Winograd, Terry. 1997. SenseMaker: An information-exploration interface supporting the contextual evolution of a user's interests. In *Proceedings of the ACM Conference on Human Factors in Computing Systems* (CHI '97), Atlanta, April, pp.11–18. New York: ACM Press. Also available on-line at http://www-diglib.stanford.edu/cgi-bin/WP/get/SIDL-WP–1996–0048.

Belkin, N. J., and Croft, W. B. 1992. Information filtering and information retrieval: Two sides of the same coin? *Communications of the ACM* 35, no. 12 (December): 29–38.

Cardie, Claire. 1997. Empirical methods in information extraction. *AI Magazine* 18, no. 4 (Winter): 65–79.

Engelbart, Douglas. 1962. *Augmenting Human Intellect: A Conceptual Framework*. Available on-line, together with a bibliography of other works by Engelbart, at http://www.bootstrap.org.

Hart, Peter E., and Graham, Jamey. 1997. Query-free information retrieval. *IEEE Expert Intelligent Systems* (September/October): 32–36.

M. H. Olson, ed. 1989. *Technological Support for Work Group Collaboration*, Hillsdale, N.J.: Erlbaum Associates.

Pedersen, Jan, Cutting, Doug, and Tukey, John. 1991. Snippet search: A single phrase approach to text access. *Proceedings of the 1991 Joint Statistical Meeting.* Atlanta: American Statistics Association.

Rose, Daniel R., Mander, Richard, Oren, Tim, Ponceleón, Dulce B., Salomon, Gitt, and Wong, Yin Yin. 1993. Content awareness in a file system interface: Implementing the "pile" metaphor for organizing information, pp. 260–69. *Proceedings of ACM-SICIR*, Pittsburgh. New York: ACM Press.

Russell, D. M., Stefik, M. J., Pirolli, P., and Card, S. K. 1993. The cost structure of sensemaking, *Proceedings of INTERCHI*, Amsterdam, April. New York: ACM Press.

Stefik, Mark. 1995. *Introduction to Knowledge Systems.* San Francisco: Morgan Kaufmann.

Stefik, Mark, ed. 1996. *Internet Dreams: Archetypes, Myths, and Metaphors.* Cambridge: MIT Press.

Stefik, Mark, and Brown, John Seely. 1989. Toward portable ideas. In Olson 1989: 147–65. (Also excerpted in Stefik 1996.)

Taubes, Gary. 1998. Malarial dreams. *Discover* (March): 108–116.

Chapter 6: The Next Knowledge Medium

Ammerman, A. J., and Cavalli-Sforza, L. L. 1984. *The Neolithic Transition and the Genetics of Populations in Europe.* Princeton, N.J.: Princeton University Press.

Boyd, R., and Richerson, P. J. 1985. *Culture and the evolutionary process.* Chicago: University of Chicago Press.

Crane, D. 1972. *Invisible Colleges: Diffusion of Knowledge in Scientific Communities.* Chicago: University of Chicago Press.

Cycorp's Home page. http://www.cyc.com/

Doyle, J., and Dean, T. 1996. Strategic directions in artificial intelligence. *ACM Computing Surveys* 28, no.4 (December). Also available at http://www.medg.lcx.mit.edu/sdcr/ai/report/latest-draft.html.

Huberman, Bernardo A. 1995. The social mind. In *Origins of the Human Brain*, Jean-Pierre Changeux and Jean Chavaillon, eds. Oxford: Clarendon Press.

Huberman, Bernardo A., and Hogg, Tad. 1995. Communities of practice: Performance and evolution. *Computational and Mathematical Organization Theory* 1, no.1: 73–92.

Pratt, Vaughan. CYC REPORT. http://boole.stanford.edu/pub/cyc.report

Stefik, Mark. 1981. Planning with constraints (MOLGEN Part 1). *Artificial Intelligence* 16, no.2 (May): 111–40. (Also reprinted in *Building Blocks of Artificial Intelligence,* E. A. Feigenbaum, ed. Reading, Mass.: Addison-Wesley, 1987.)

Whitten, David. The Unofficial, Unauthorized CYC Frequently Asked Questions Information Sheet. http://www-laforia.ibp.fr/~cazenave/cyc.html

Chapter 7: The Edge of Chaos

Blinder, Paul, and Quandt, Richard. 1997. The computer and the economy. *Atlantic Monthly* (December): 26–32.

Friedlander, Amy. 1995. *Emerging Infrastructure: The Growth of Railroads.* Reston, Va.: Corporation for National Research Initiatives.

Wysocki, Bernard, Jr. 1980. Growing home-information field led by telecomputing's "Source" for news. *Wall Street Journal*, January 30, p. 8.

URLs for Erie Canal Sources

There are several on-line sources of information about the Erie Canal.

New York State Canal System Home Page. Links to many other canal-related sites. http://www.thruway.state.ny.us/canals/index.htm.

Erie Canal Museum, Syracuse, N.Y. http://www.cnyric.org/cnyregion/canal/ and http:/www.syracuse.com/host/eriecanal/index.html

Brief history of the Erie Canal in Buffalo, N.Y. http://intotem.buffnet.net/bhw/eriecanal/ditch/ditch.htm.

Chapter 8: The Digital Keyhole

Abelson, H., Anderson, R., Bellovin, S. M., Benaloh, J., Blaze, M., Diffie, W., Gilmore, J., Neumann, P. G., Rivest, R. L., Schiller, J. I., Schneier, B. *The Risks of Key Recovery, Key Escrow, and Trusted Third-Party Encryption.* Available on-line at http://www.crypto.com/key_study/.

Agre, Philip E., and Rotenberg, Marc. eds. 1997. *Technology and Privacy: The New Landscape*, Cambridge: MIT Press.

Bennett, Colin J. *Regulating Privacy: Data Protection and Public Policy in Europe and the United States,* Cornell University Press, 1992.

Cool, C., Fish, R., Kraut, R., Lowery, C. 1992. Iterative design of video communication systems. *Proceedings of the ACM Conference on Computer-Supported Cooperative Work*, Toronto. New York: ACM Press.

Denning, D., and Branstad, D. 1996. A taxonomy for key escrow encryption systems. *Communications of the ACM* 39, no. 3: 34–40.

Gellman, Robert. Does privacy law work? In Agre and Rotenberg 1997.

Harrison, S., Bly, S., Minneman, S., and Irwin, S. 1997. The media space. In *Video-mediated Communication,* K. Finn, ed. Hillsdale, N.J.: Erlbaum Associates.

Lewis, Peter H. 1998. Forget Big Brother. *New York Times*. March 19. Available on-line at http://www.nytimes.com/library/tech/yr/mo/circuits/articles/19data.html.

Michalski, Jerry. 1998. Identity management. *Release* 1 (February): 23.

Patton, Phil. Caught: You used to watch television. Now it watches you. *Wired.* Available on-line at http://www.wired.com/wired/3.01/features/caught.html.

Stults, R. 1988. *The Experimental Use of Video to Support Design Activities.* Report SSL–89–19. Palo Alto: Xerox PARC.

Want, R., Hopper, A., Falcco, V., and Gibbons, J. 1992. The active badge location sytem. *ACM Transactions on Office Information Systems* 10, no. 1: 91–102.

URLs for Legislation and Government Documents on Privacy and Security

British Columbia Provincial Government. Freedom of Information and Protection of Privacy Act. Available at http://www.grannyg.bc.ca/FIPA/law/overview.html.

Government of Canada. Consultation document on privacy. Available at http://strategis.ic.gc.ca/privacy/.

Council of Europe. 1981. *Convention for theProtection of Individuals with Regard to Automatic Processing of Personal Data*,. Available on-line at http:/www2.echo.lu/legal/en/dataprot/counceur/conv.html and http://www.unesco.org/webworld/com/compendium/3205.html.

National Information Infrastructure Task Force (U.S.). 1995. Privacy Working Group, Information Policy Committee. *Privacy and the National Information Infrastructure: Principles for Providing and Using Personal Information.* June 6. Available on-line at http://www.iitf.nist.gov/ipc/ipc/ipc-pubs/niiprivprin_final.html.

National Information Infrastructure Task Force (U.S.). 1997. *Options for Promoting Privacy on the National Information Infrastructure.* April. This extensive review of privacy options focuses on four critical areas of privacy: federal government records, personal communications, medical records, and the marketplace. It is available on-line at http://www.iitf.nist.gov/ipc/privacy.htm.

National Institute of Standards and Technology (U.S.), Computer Security Resource Clearinghouse. Key recovery information and notes from a technical advisory committee on key recovery. Available at http://csrc.ncsl.nist.gov/keyrecovery/ and at http://csrc.ncsl.nist.gov/tacdfipsfkmi/.

Organisation of Economic Cooperation and Development. 1980. Guidelines on Protection of Privacy and Transborder Data Flow. Available on-line at http://www.oecd.org/dsti/sti/it/secur/prod/.

United States Constitution and index to relevant cases. http://www.law.emory.edu/FEDERAL/usconst.html and http://www.access.gpo.gov/congress/senate/constitution/

URLs for Organizations Concerned with Privacy and Security

Center for Democracy and Technology. Its web page demonstrates issues of privacy by retrieving information about you from your browser. Available at http://www.13x.com/cgi-bin/cdt/snoop.pl.

Computers Freedom and Privacy Conference. Information about this conference held annually in the United States available at http://www.cfp.org/.

Electronic Privacy Information Center (EPIC). The web page of this public interest research center located in Washington, D.C., provides a guide to many Internet privacy resources. Available at http://www.epic.org/privacy/privacy_resources_faq.html.

Information Industry Technology Council. IITC, an organization of U.S. providers of information technology products and services, was founded in 1916. Its principles for privacy and data protection can be seen on-line at http://www2.itic.org/itic/iss_pol/ppdocs/pp_privprin.html.

Open Profiling Standard. Proposed standard for a uniform architecture to manage personal information at Internet sites. Information available at http://developer.netscape.com/ops/ops.html.

Privacy International. PI, a human rights group based in London, was formed in 1990. Its web page reporting on international privacy news is available at http://www.privacy.org/pi/.

TRUSTe. An independent, nonprofit organization dedicated to establishing a trusting environment on the Net. Available on-line at http://www.truste.org/users/abouttruste.html.

World Wide Web Consortium (W3C). Architecture Working Group. Platform for Privacy Preferences (P3P). Working documents providing an overview of a P3P architecture and a grammar for negotiation. Available on-line at at http://www.w3.org/TR/WD-P3P-arch.html and http://www.w3.org/TR/WD-P3P-grammar.html.

Chapter 9: Strangers in the Net

Hegener, Michiel. 1996. The internet, satellites, and economic development. *On The Internet* 2, no. 5 (September/October). Available on-line at http://www.isoc.org/isoc/publications/oti/articles/theinternet.html.

Kahin, Brian, and Keller, James. 1996. *Public Access to the Internet.* Cambridge: MIT Press.

Kahin, Brian, and Nesson, Charles, eds. 1997. *Borders in Cyberspace: Information Policy and the Global Information Infrastructure.* Cambridge: MIT Press.

Macy, Joanna. 1991. *World as Lover, World as Self.* Berkeley, Calif.: Parallax Press.

Mander, Jerry, and Goldsmith, Edward, eds. 1997. *The Case Against the Global Economy: And for a Turn Toward the Local.* San Francisco: Sierra Club Books.

Meena, C. K. 1994. The bane of cultural pollution. *The Hindu,* May 1. Available on the Net at http://www.prakash.org/experience/places/touhin19940501_00.html.

Simons, Marlise. 1997. A delicate seaweed is now a monster of the deep. *New York Times.* August 16. Available on-line at http://www.nysaes.cornell.edu/ipmnet/archive/med.seaweed.html.

Skolnikoff, Eugene. 1993. *The Elusive Transformation: Science, Technology, and the Evolution of International Politics.* Princeton, N.J.: Princeton University Press.

Walsh, Roger, and Vaughan, Frances, eds. 1993. *Paths Beyond Ego*. Los Angeles: Tarcher/Perigee.

URLs for Internet Governance and history

At the time of this writing, thousands of organizations are participating in some way in governing the Internet and in creating and implementing standards for it. A query from a Net search engine for pages containing the words *Internet Society* yields thousands of sites for chapters and branches in countries, states, and regions around the world. Here are just a few of them.

Internet Engineering Task Force (IETF) Home Page. http://www.ietf.cnri.reston.va.us/home.html

Internet History. A page with links to several accounts is available at http://www.isoc.org/internet/history/.

Internet Society. (ISOC) Home Page. http://www.isoc.org/

ISOC page on internationalization of the Internet. Special focus on multilingual web pages. http://www.isoc.org:8080/

Network Business Services Home Page. With links to the Internet Architecture Board, the Internet Engineering Task Force, World Wide Web Consortium, and many other bodies concerned with and participating in evolution of the Net. http://www.nim.com.au/inet_pub/in04001.htm

Web page on efforts to regulate the Internet in Singapore. Contains a list of articles. http://www.fl.asn.au/singapore/refs.html

World Intellectual Property Organization (WIPO) Home Page. http://www.wipo.org/

World Wide Web Consortium (W3C) Home Page. http://www.w3.org/

Chapter 10: Indistinguishable from Magic

Amato, Ivan. 1992. Animating the material world. *Science* 155 (January 17) :5042.

Biddle, Frederic M., Lippman, John, and Mehta, Stephanie. 1998. One satellite fails, and the world goes awry. *Wall Street Journal*, Thursday, May 21, p. B1.

Bly, Robert. 1988. *A Little Book on the Human Shadow*. San Francisco: Harper.

Clarke, Arthur C. 1945. Extra-terrestrial relays. *Wireless World* (October): 305–309. Available as an on-line image at http://www.lsi.usp.br/~rbianchi/clarke/ACC.ETRelaysFull.html and as digital text at http://alto.histech.rwth-aachen.de/www/leute/friedewald/clarke/clarkeextra.html.

Hinden, Robert M. IP next generation overview. Available on-line at http://playground.sun.com/pub/ipng/html/INET-IPng-Paper.html.

URLs for Wearable Computers

At the time of this writing, the technology of wearable computers is novel, immature, and rapidly evolving. The following URLs are representative but may have a relatively short half-life on the Net. The first four are a sampler of research projects in wearable computing currently being conducted at universities in the United States.

Carnegie-Mellon University Wearable Computer Systems. http://almond.srv.cs.cmu.edu/afs/cs/project/vuman/www/index.html

Columbia University Augmented Reality Projects. http://www.cs.columbia.edu/graphics/

Georgia Institute of Technology Wearables Project. http://mime1.marc.gatech.edu/EPSS/

MIT Wearables Project. Also provides a page of links to other relevant Net sites. http://wearables.www.media.mit.edu/projects/wearables/ and http://wearables.www.media.mit.edu/projects/wearables/wearlinks.html

The next group of URLs are a representative sampler of pages about commercially developed wearable systems.

Dallas Semiconductor. The iButton and JavaRing. http://www.ibutton.com and http://www.idg.net/idg_frames/english/content.cgi?vc=docid_9%2d31299%2ehtml

Defense Advance Research Projects Agency (DARPA). Web page describing the Warfighter Visualization Program. http://web-ext2.darpa.mil/ETO/wv/index.html

MicroOptical. Normal looking eyeglasses with a built-in computer monitor. http://www.microopticalcorp.com/eyefaq.htm

Phoenix Group. Descriptions of wide range of wearable computer products. http://ivpgi.com/

Rockwell International. Information about the Trekker guidance system. http://www.cacd.rockwell.com/gen_info/news/releases/cacd8095.htm

Symbol Technologies. A radio-linked wrist computer with a scanner ring. http://www.symbol.com/ST000262.htm

Virtual Visions. A system combining a speech interface with a head-mounted display. http://www.hypervision.co.uk/speech/vvision.htm

URLs for Smart Buildings and Appliances

Technology for smart buildings and smart homes has been around for several years. Here are some on-line sources that refer to it.

CEBus/PnP. A general description of the common application language and a fact sheet about PnP can be found on-line, respectively, at http://www.cebus.org/10.htm and http://www.cebus.org/frmain.htm#FAQ_1.2.

Echelon Corporation. On-line information about their LonWorks technology is av ailable at http://www.hometoys.com/htinews/dec96/articles/lonworks.htm and http://www.apspg.com/products/mdad/lonworks.html.

IPv6 and IPng Internet protocols. Information about developments in these protocols are available on-line at http://playground.sun.com/pub/ipng/html/ipng-main.html and http://playground.sun.com/pub/ipng/html/INET-IPng-Paper.html.

Kitchen Sync Project at the MIT Media Lab. Information available at http://www.media.mit.edu/pia/Pubs/AprilDemo/kitchensync/.

Smart House Consortium Home Page. http://www.smart-house.com

X–10 System. Page describing components from several manufacturers. http://web.cs.ualberta.ca/~wade/HyperHome/Faq/faq_section2.html

Michael Schrage's fictional account of his experiences with his SmartHouse originally appeared in the *Los Angeles Times* in 1993. When I checked with a search service, I found variants of it at about 100 places on the Net. Here's one of them: http://www.shults.com/hypermail/jokes/0007.html.

On-line toaster jokes have also made their way around on the Net. Here's one source: http://web.shorty.com/geeks/96/dec/msg00000.html.

URLs for Stereolithography and Manufacturing

Brian Curless and Marc Levoy, Stanford University. Describes building a 3-D fax machine by combining range-scanning technology with stereolithography. http://www-graphics.stanford.edu/projects/faxing/happy/

Modern Machine Shop. A professional on-line magazine about trends and technologies in automated manufacturing. http://www.mmsonline.com

University of Utah. Department of Mechanical Engineering Home Page. Information on rapid-prototyping technology. Includes extensive lists of commercial systems for rapid prototyping, concept modellers, venders, service providers. and an on-going record of research projects, literature, and conferences around the world. http://stress.mech.utah.edu/home/novac/rapid.html

Epilogue

Hodder, Ian. 1982. *Symbols in Action. Ethnoarchaeological Studies of Material Culture*. Cambridge, England: Cambridge University Press.

Credits

Chapter 3, "The Bit and the Pendulum" is based on a paper by Mark Stefik and Alex Silverman that appeared in the September 1997 issue of *American Programmer*.

Chapter 4, "The Digital Wallet and the Copyright Box" is largely based on a working paper by Mark Stefik and Xin Wang.

Chapter 6, "The Next Knowledge Medium" is an updated version of an article of the same name published in *AI Magazine* 7, no. 1 (Spring 1986): 34–46.

Index